UNDERSTANDING NE CODE RULES ON

GROUNDING & BONDING

BASED ON THE 1999 NE CODE

2ND EDITION

By JOHN PASCHAL, P.E.
Edited by FRED HARTWELL

A PRIMEDIA Intertec
Electrical Group Publication
Overland Park, KS 66212-2216

UNDERSTANDING NE CODE RULES ON GROUNDING & BONDING, 2ND EDITION
BASED ON THE 1999 NE CODE
© 1998 PRIMEDIA Intertec

1st Printing, September, 1998

Published by
EC&M Books
PRIMEDIA Intertec
9800 Metcalf Ave.
Overland Park, KS 66212–2216

ISBN 0-87288-694-8
Library of Congress Card Catalog Number: 98-072568

Please note: The designations "National Electrical Code," "NE Code," and "NEC," where used in this book, refer to the National Electrical Code®, which is a registered trademark of the National Fire Protection Association.

CONTENTS

PREFACE

GROUNDING AND BONDING, as spelled out in Article 250, is one of the most important subjects covered by the NEC. Its purpose is to assure the safety of persons working on or around electrical equipment and installations. It is, however, a very complex subject and the set of rules dealing with grounding and bonding reflect this fact. The purpose of this book is to help those who must apply the rules to fully understand the meaning of these regulations and how they are to be used to prepare the design of an electrical system.

All of the sections of Article 250 are covered in this book. They have been rewritten in descriptive, rather than mandatory, language. Thus, the word "shall" has been changed to "must" or some similar word or phrase. The sections and subsections have, in addition, been expanded to include explanations of what is intended. Where sections of other articles in the Code appear as part of the NEC text, the citation has been included as part of the description of the rule's intent. As often as possible, illustrations have been integrated to give a better picture of the meaning of the rule and how it is to be carried out.

The order in which the subject appears in this book does not totally follow the sequence in which the parts and sections appear in the NEC to treat central safety concepts first before the rules applying them are covered.

One additional change in sequence has been made. Rules that deal with the sizing of the various components of a grounding system appear in many of the various sections of Article 250. This makes for a lot of page turning to refer to tables and other rules. To simplify the procedure, all references to sizing have been accumulated in Chapter 11 of this book. The cited tables appear there along with examples of the calculations involved in a typical installation.

Editors of EC&M

ART. 100 & Art. 250, Part A — DEFINITIONS AND BASIC PRINCIPLES

Grounding and bonding are among the most important topics covered in the National Electrical Code. They are essential for the safety of personnel working around electrical equipment, as well as for proper operation of circuit protective devices. Also, proper grounding and bonding techniques are essential in assuring that sensitive electronic equipment is protected from transients and other spurious signals that can seriously affect the way data is processed.

Ground is defined in the NEC Article 100, "Definitions," as:

A conducting connection, whether intentional or accidental, between an electrical circuit or equipment and the earth, or to some conducting body that serves in place of the earth.

Given this definition, the word "ground" cannot be used interchangeably with "grounded" or "grounding". A ground describes the state in which two conductive objects are in contact with each other and/or the earth. In the electrical industry, the standard symbol used on one-line drawings to represent a ground is shown in **Fig. 1.1**.

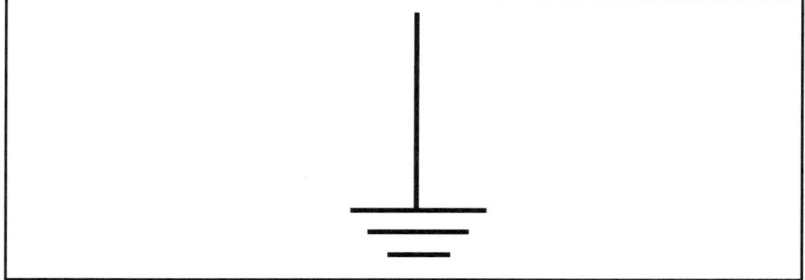

Fig. 1.1 This is the standard symbol for a *ground*. Any system or equipment that is shown connected to this symbol is considered to be *grounded*.

On the other hand, **Grounded** is defined as:

Connected to earth or to some conducting body that serves in place of the earth.

This defines an object that has been attached, either purposefully or accidentally, to the earth or some object. For instance, a ground bus of a switchgear is an adequate place to which to terminate an equipment grounding conductor for purposes of grounding, because the grounding bus is connected to the earth.

Further definitions are given in Article 100 that help clarify the proper use of words.

Effectively Grounded means:

Intentionally connected to earth through a ground connection or connections of sufficiently low impedance and having sufficient current-carrying capacity to prevent the buildup of voltages that may result in undue hazards to connected equipment or to persons.

Grounded Conductor describes the following:

A system or circuit conductor that is intentionally grounded.

From these definitions it is clear that it is wrong to say, "It is a ground conductor." What must actually be said is: "It is a grounded conductor," "It is an equipment grounding conductor," or "It is a grounding electrode conductor."

Grounding is a word that is not defined in the NEC, but is used in the definitions in Article 100 as a descriptive term. For instance, a Grounding Conductor is described as:

A conductor used to connect equipment or the grounded circuit of a wiring system to a grounding electrode conductor or to grounding electrodes.

Also, an **Equipment Grounding Conductor** is defined as being:

The conductor used to connect the noncurrent-carrying metal parts of equipment, raceways, and other enclosures to the system grounded conductor, the grounding electrode conductor, or both, at the service equipment or at the source of a separately derived system.

In addition, a **Grounding Electrode Conductor** is defined as :

The conductor used to connect the grounding electrode to the equipment grounding conductor, to the grounded conductor, or to both, of the circuit at the service equipment or at the source of a separately derived system.

Fig. 1.2 graphically shows the relationship between these three different conductors and visually clarifies the location and use of each. Note that under present definitions, grounding electrode conductors only exist at a service entrance. Conductors serving a similar function but connected downstream in the system (other than at separately derived systems), such as at a building disconnect at a building served from another building, are properly termed grounding conductors.

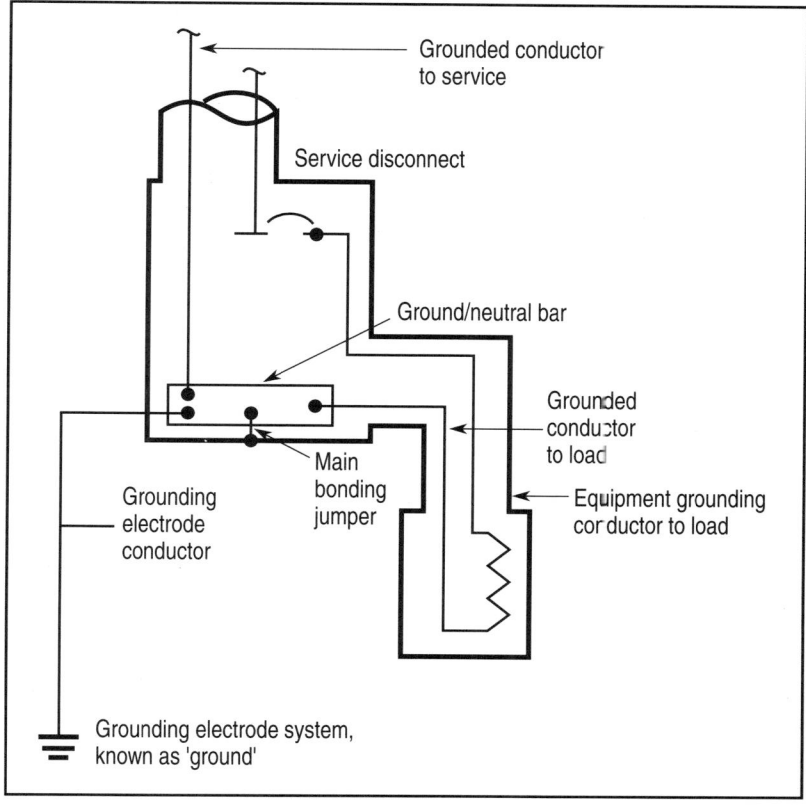

Fig. 1.2 Grounded, grounding, and grounding electrode conductors.

SEC. 250-2 – TYPES OF GROUNDING

There are two distinct types of grounding required by the NEC: system grounding and equipment grounding. It is important to understand the difference in function between the two.

System grounding. Fig. 1.3 illustrates one reason certain systems must be grounded and one of the most important benefits of doing so. Sec. 250-2(a) of the NEC points to these performance objectives:

(a) Grounding of Electrical Systems. Electrical systems that are required to be grounded shall be connected to earth in a manner that will limit voltages imposed by lightning, line surges, or unintentional contact with higher voltage lines, and that will stabilize the voltage to earth during normal operation.

In effect, system grounding usually refers to whether the secondary windings of a transformer are grounded, although a system derived from any source can be grounded. In an *ungrounded* delta-connected 3-phase

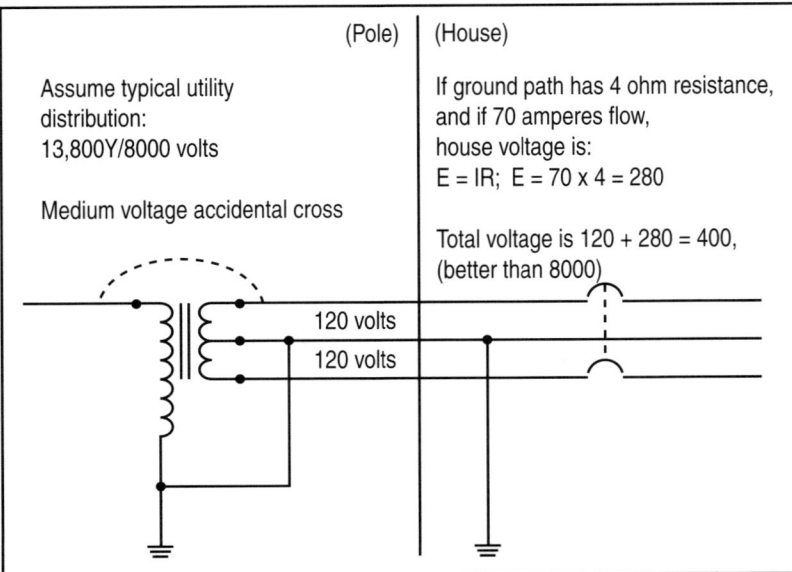

Fig. 1.3 This grounded system greatly reduces the voltage exposure from the utility primary conductors.

system shown in **Fig. 1.4(A)**, single phase-to-ground faults produce very low values of fault current. These currents are not sufficient for the operation of overcurrent relaying whose purpose is to isolate the faulted circuits to prevent arcing and fire damage.

A *grounded* wye-connected 3-phase system, shown in **Fig 1.4(B)**, has its center point connected to ground. Such a system permits automatic clearing of accidental grounds because phase-to-ground faults produce currents that can be used with overcurrent relaying. Another advantage

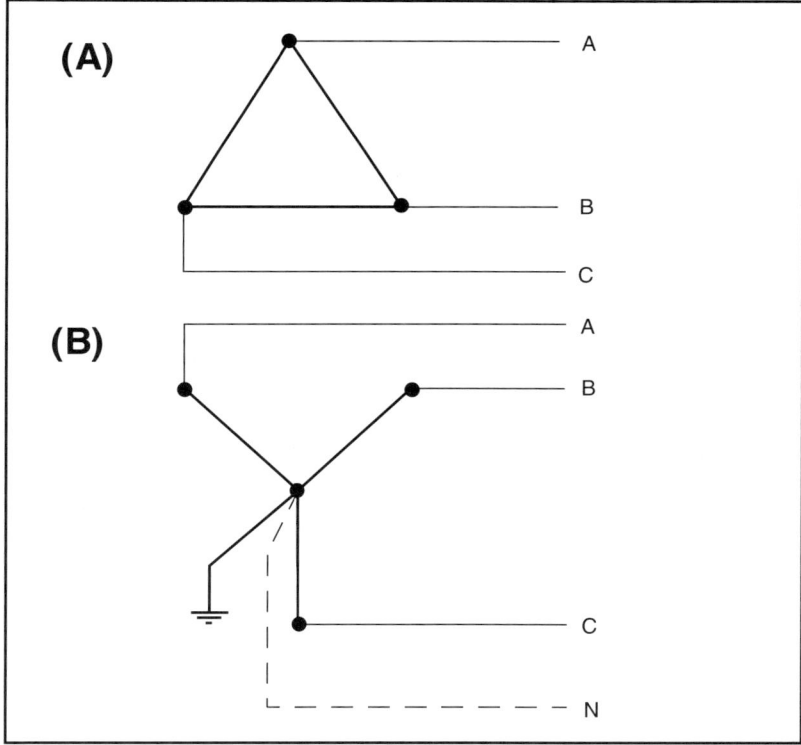

Fig. 1.4 System grounding refers to whether the secondary windings of a transformer are grounded. In the 3-phase delta-connected secondary shown in **(A)**, there is no connection the ground. The 3-phase wye-connected secondary shown in **(B)**, however, has its midpoint connected to ground. Single-phase systems can also be grounded or ungrounded.

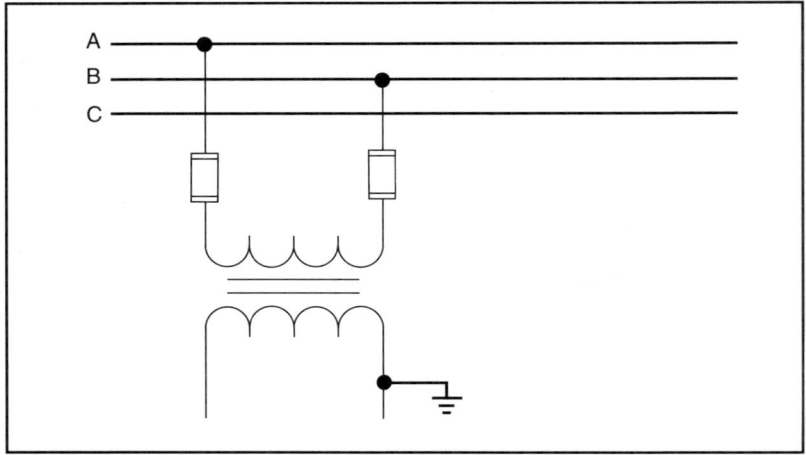

Fig. 1.5 The electrical system can also be grounded at other points of the distribution system within a facility, principally at the secondaries of transformers that further reduce the in-plant distribution voltage. These are called separately derived systems in the code.

of such a system is that a conductor from the grounded center point (a neutral) can be carried with the ungrounded conductors and used in the distribution system for connecting loads at a lower voltage from phase to neutral.

It is also possible to create a grounded system from a delta-connected secondary of a transformer All that must be done is to ground one of the three phases derived from the transformer.

A grounded system can also be created further downstream within a power distribution system, mainly at transformers used to step down voltage. These connections to ground (**Fig. 1.5**) are simply a variation of a grounded system as defined in the NE Code.

Equipment grounding. Grounding of equipment has a quite different purpose, as described in Sec. 250-2(b), as follows:

(b) Grounding of Electrical Equipment. Conductive materials enclosing electrical conductors or equipment, or forming part of such equipment, shall be connected to earth so as to limit the voltage to ground on these materials. Where the electrical system is required to be grounded, these materials shall be connected together and to the supply system grounded conductor as specified by this article. Where the electrical

system is not solidly grounded, these materials shall be connected to-
gether in a manner that establishes an effective path for fault current.

In grounded systems, equipment grounding conductors are bonded to
the system grounded conductor to provide a low impedance path for
fault current that will facilitate the operation of overcurrent devices un-

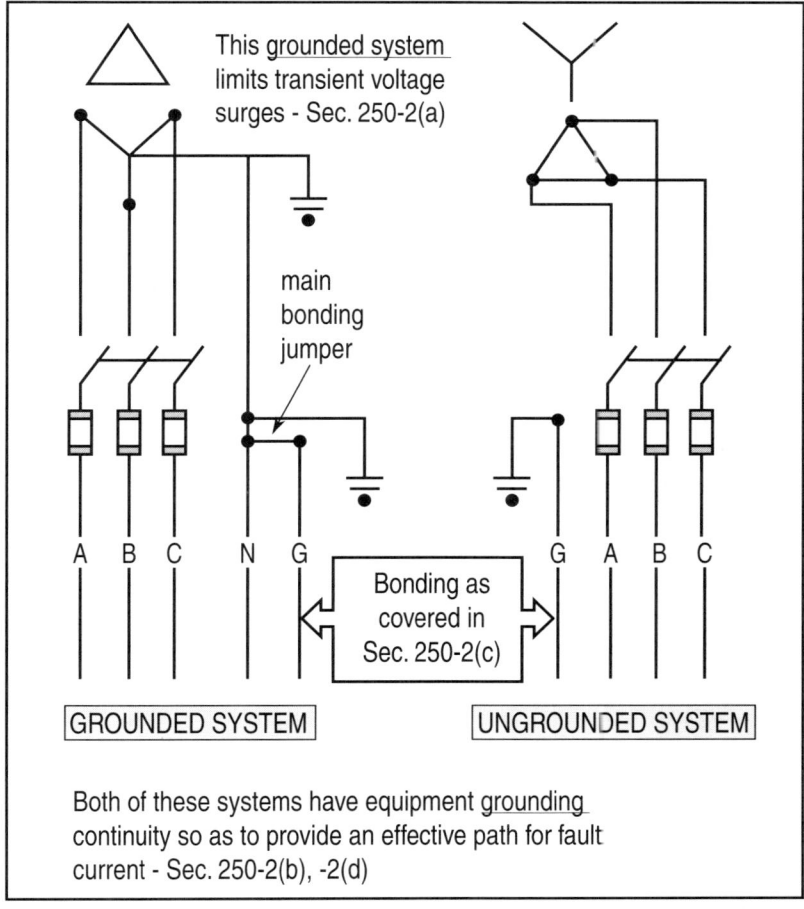

This grounded system
limits transient voltage
surges - Sec. 250-2(a)

main
bonding
jumper

A   B   C   N   G

Bonding as
covered in
Sec. 250-2(c)

G   A   B   C

GROUNDED SYSTEM

UNGROUNDED SYSTEM

Both of these systems have equipment grounding
continuity so as to provide an effective path for fault
current - Sec. 250-2(b), -2(d)

Fig. 1.6 A fault current path may, or may not be a fault current *return* path, depending
on whether the system is, or is not grounded.

der ground-fault conditions. Note that the wording of neither (b) above nor (d) below uses the familiar terminology "fault current return path." The reason is that this material covers both grounded and ungrounded systems (**Fig. 1.6**).

Ungrounded systems, by definition, have no point to return fault current to. Nevertheless, the fault current path must be reliable, because if two simultaneous faults occur from different phases on an ungrounded system, the equipment grounding system, as it exists between those two points, must safely carry the resulting phase-to-phase fault current.

The first paragraph of Sec. 250-2(d) states:

> **(d) Performance of Fault Current Path.** The fault current path shall be permanent and electrically continuous, shall be capable of safely carrying the maximum fault likely to be imposed on it and shall have sufficiently low impedance to facilitate the operation of overcurrent devices under fault conditions.

As described, equipment grounding involves connecting metallic enclosures, raceways, motor frames, building steel, and other items that may accidentally become energized, to ground to maintain a low potential difference between conductive items that are close to each other. This is done to prevent a person from receiving a shock if two items are touched simultaneously, such as touching a motor housing while standing on a concrete floor.

For example, as shown in **Fig. 1.7(A)**, a metallic enclosure grounded only through the entering conduit contains a faulted phase conductor that is touching one of the surfaces. If the grounding path back to the source provided by the conduit is poor (due to corrosion, etc.) or does not exist, and a person simultaneously contacts the enclosure and a grounded surface, some of the fault current will flow through the person. On the other hand, as shown in **Fig 1.7(B)**, if the enclosure has a solid connection to ground, the resistance through the person's body will be considerably higher than that of the ground return path. In this event, only a very small percentage of the fault current will flow through the person. In addition, the low impedance in the ground return path means that a high value of fault current will flow, causing the overcurrent protective device to clear the fault quickly. **Note that if the equipment grounding conductor is of essentially zero impedance, then the enclo-**

sure the person is touching and the surface the person is standing upon will have no potential difference between them, and essentially no current will flow through the person. This is a principal human safety objective of Article 250.

The earth is not an equipment grounding conductor. Although we fondly speak of grounding in terms of safety, and think of the earth, one of the great ironies is that the earth is nearly the most unreliable equip-

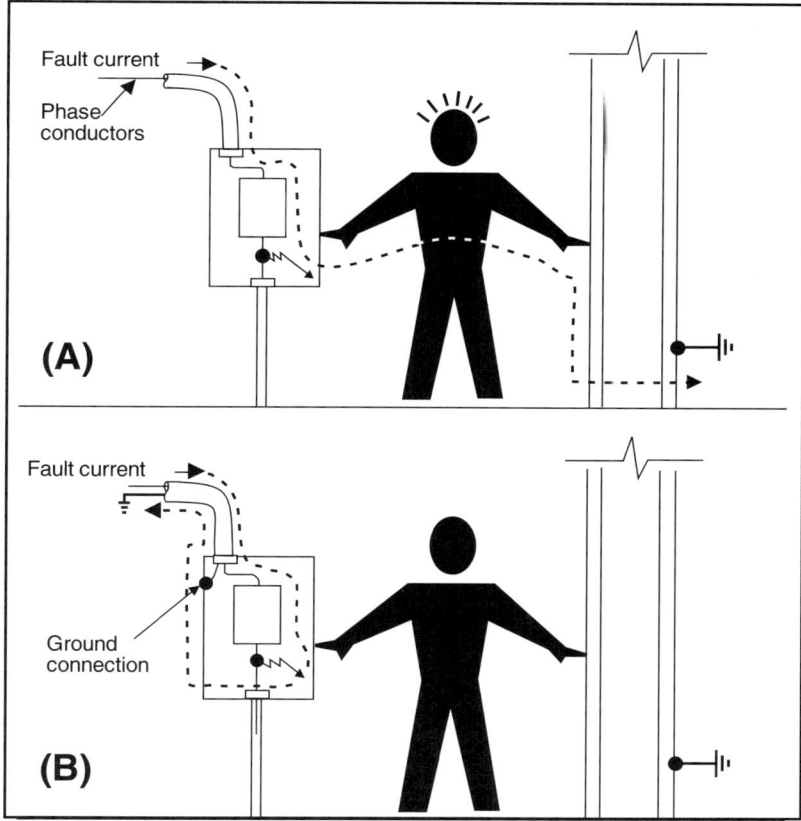

Fig. 1.7 An ungrounded or poorly grounded metallic enclosure of a piece of electrical equipment can allow a hazardous amount of fault current to flow through the body of a person contacting it and a grounded surface simultaneous **(A)**. This shock hazard can be reduced by solidly grounding the enclosure **(B)**.

ment grounding conductors imaginable. This is because its functional conductivity is effectively limited by the resistance of whatever electrode is connected to it. The very best electrodes have resistances in the one-ohm range; through it, on a 120 V circuit a fault into the earth would only pass 120A, by Ohm's Law.

That amount of current would never trip a 150A overcurrent device, and it would be below the instantaneous tripping points of lower rated protective devices. That is, even a 30A device would see the 120A as an overload, such as a motor attempting to start. As such, it would not trip promptly. Remember, typical grounding resistances to earth run much higher than 1 ohm, and you see why the final paragraph of Sec. 250-2(d) reads as it does:

> *The earth shall not be used as the sole equipment grounding conductor or fault current path.*

Make as many supplementary connections to earth from the equipment grounding conductor(s) as you wish, but make absolutely certain that a code-compliant equipment grounding conductor is in place to carry any fault current that may be imposed. This concept is reiterated in Sec. 250-54. See **Fig. 1.8**.

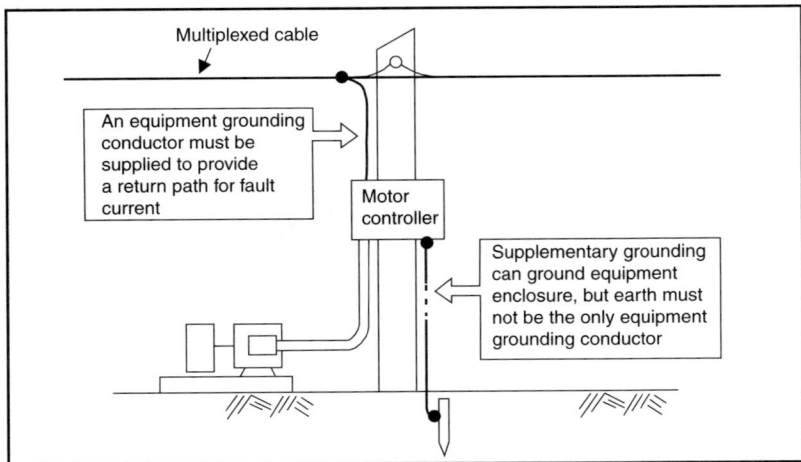

Fig. 1.8 The cable or raceway to remotely located equipment must contain an equipment grounding conductor. The equipment can also, in addition be grounded locally.

BONDING TERMS

Grounding and bonding are often used in place of each other when describing grounding; but the two terms are not synonymous, nor can they be used interchangeably.

Bonding is defined in Article 100 as being:

> *The permanent joining of metallic parts to form an electrically conductive path that will assure electrical continuity and the capacity to conduct safely any current likely to be imposed.*

Bonding, thus is the means by which effective grounding can be achieved. Bonding electrical equipment, metal raceways, and enclosures provides a continuous equipment grounding conductor and an effective low-impedance path through which short circuit currents to ground can flow. This assures that protective devices in the affected circuit will trip and "open" quickly. Also, a continuous equipment grounding conductor assures that short circuit current flow will not cause arcing and sparking along such a path. This is particularly important in hazardous locations.

One way in which bonding can be accomplished is via a **Bonding Jumper**, which is defined as being:

> *A reliable conductor to assure the required electrical conductivity between metal parts required to be electrically connected.*

The types of bonding jumpers are defined in Article 100. They will be discussed further in Chapter 3 of this book.

The term **Equipment Bonding Jumper** describes:

> *The connection between two or more portions of the equipment grounding conductor.*

A **Main Bonding Jumper** is defined as being:

> *The connection between the grounded circuit conductor and the equipment grounding conductor at the service.*

In addition to bonding within an electrical equipment grounding path, other building systems may need to be bonded to the electrical system for safety reasons, as covered in Sec. 250-2(c):

> **(c) Bonding of Electrically Conductive Materials and Other Equipment.** *Electrically conductive materials, such as metal water piping, metal gas piping, and structural steel members, that are*

likely to become energized shall be bonded as specified by this article to the supply system grounded conductor or, in the case of an ungrounded electrical system, to the electrical system grounded equipment, in a manner that establishes an effective path for fault current.

The bonding requirements in Art. 250 are covered in Chapter 3 of this book.

Some perspective for the reader on Art. 250 and the NEC in general:

The NEC is a prescriptive, not a performance-based Code. That is, it doesn't tell us what electrical protective protocols are supposed to accomplish, it tells us instead what we are to do to accomplish the safety objectives in Sec. 90-1. The prescriptive requirements in the NEC result from a consensus process whereby the collective wisdom of international experience gets boiled down to actual requirements. If you want to know what a performance based code would look like, imagine Art. 250 with just Sec. 250-2. Everything else is essentially prescriptive.

Although performance-based codes are all the rage, they require incredibly sophisticated engineering support and inspection to work. Many times inspectors have seen engineered plans that failed to comply with provisions of this or other articles. Often the engineer will respond with the excuse of equivalent safety. However, if the real reason for the requirement were known, it would have been obvious that the proposal was not equivalently safe. There simply isn't any substitute for the collective wisdom of generations of participants in the resulting document.

That said, there also isn't any substitute for good theoretical knowledge of the reasons that underlie those requirements. It allows us to be multidimensional in outlook, going beyond "it says do it like this so I'm doing it like this." There isn't any aspect covered by the NEC for which that is more true than grounding. If you understand, *really understand everything in this section*, then everything that follows should make sense to you. After you read this book, come back and reread this chapter, and be sure.

SEC. 250-6 – OBJECTIONABLE CURRENT OVER GROUNDING CONDUCTORS

Speaking of sophissticated engineering support and inspection, this section must be used with care because it almost seems to give blanket authority to do whatever is necessary to stop objectionable currents from flowing in the grounding system. Such is not the intent. As a matter of fact, this section specifically excludes tampering with the grounding system to overcome "noise" that may be causing difficulties in sensitive electronic equipment [see subsection (d) below].

The permission granted in this section principally deals with objectionable currents that can flow over grounding conductors due to a circuit containing severely unbalanced loads or due to improper installation practices. Because of the complexity of most grounding systems and the great number of interconnections between equipment and structures in such a system, the NEC allows modifications to be made to the grounding system and connections to overcome such problems. Those permitted are detailed in the subsections that follow.

(a) Arrangement to prevent objectionable current. Grounding of electric systems, circuit conductors, surge arresters, and conductive non-current-carrying materials and equipment must be installed and arranged in a manner that will prevent an objectionable flow of current over the grounding conductors or grounding paths.

It is the responsibility of those designing and installing a grounding system to assure that "single-point grounding" and other sound concepts, as well as the requirements of Article 250, are met. A well-planned grounding and power distribution system will probably satisfy all needs.

(b) Alterations to stop objectionable current. If the use of multiple grounding connections results in an objectionable flow of current, one or more of the following alterations are permitted to be made, provided that the requirements of Sec. 250-2(d) , which describes an effective grounding path, are met. The permitted alterations are:

• discontinue one or more, but not all, of the grounding connections;
• change the locations of the grounding connectors;
• interrupt the continuity of the conductor or conductive path interconnecting the grounding connections; and/or
• take other suitable remedial action satisfactory to the authority having jurisdiction.

(c) Temporary currents not classified as objectionable currents are those resulting from accidental conditions, such as ground-fault currents, that occur only while the grounding conductors are performing their intended protective functions. This, however, does not exclude making changes in the system to correct excessive flow of current in the grounding system during a fault when that condition is due to improper connections, etc. in the grounding system. That is, the Code expects improper connections to be corrected.

(d) Limitations to permissible alterations. The intent of Sec. 250-58 *is not* that of permitting electronic equipment to be operated on AC systems or from branch circuits that are not grounded as required by Article 250. Currents that introduce noise or data errors in electronic equipment also *are not* considered to be the objectionable currents addressed in this section.

Grounding remains as much as art as a science. Theoretically all grounding is referenced to ground (the earth), and thus, there should be zero voltage difference between any two grounded points within an electrical system. In actuality, however, voltage differences do exist because impedances to ground are not equal throughout a grounding system, the resistance of the earth varies according to its composition, poor or improper connections may exist, and many other possibilities may cause

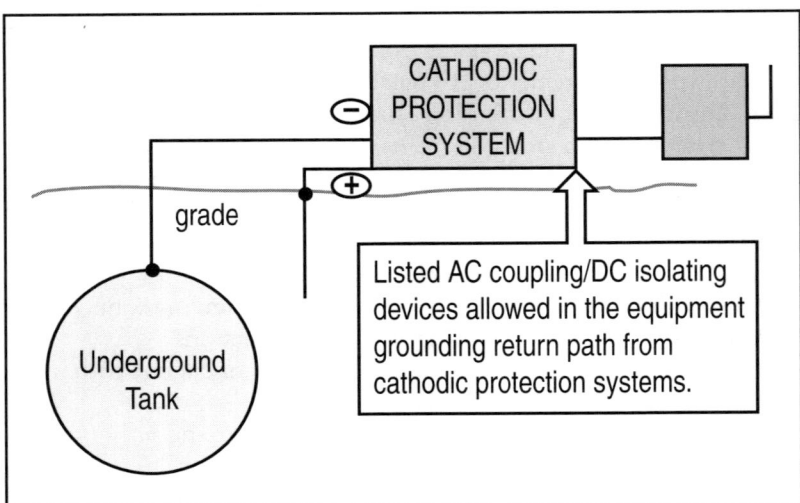

Fig. 1.9 This is a special application, addressing objectionable DC currents over grounding conductors.

unequal ground current flow and voltage differences.

Voltage differences allow unwanted currents to flow in the grounding conductors. Also, induced noise may also travel over this path. Because of the great number of interconnections that are likely to exist in a grounding system, some of these currents can enter circuitry of sensitive electronic equipment and cause upsets. This, however, is not to be used as a reason to disconnect or otherwise tamper with the grounding connections. Once again, it is the responsibility of those designing and installing the power and grounding system to provide a reliable system to support this type of equipment.

(e) **Isolation of objectionable dc ground currents**. You can use a listed ac coupling/dc isolation device to prevent damage from cathodic protection systems (see **Fig. 1.9**).

There are devices that block dc current, while allowing an ac fault to pass, allowing an overcurrent device to operate and preventing hazardous voltages from remaining on conductive surfaces. These devices and this new allowance should allow you to avoid the temptation of discontinuing the equipment grounding return path from a cathodic protection source because of dc currents traveling over those pathways.

These protection systems inject low levels of dc current into the ground, arranged so the object being protected is a cathode. As such, the only item that corrodes is the anode, which is sacrificial and readily replaceable as needed. As long as the dc stays in the ground, there isn't any problem. However, it can also show up in the ac side of the system because of the necessary continuity on the equipment grounding return path. Routine dc currents on electrical raceways, etc. can cause serious corrosion and other problems.

ARTICLE 250, PART C — GROUNDING ELECTRODE SYSTEM AND GROUNDING ELECTRODE CONDUCTOR

2

To ground an electrical system, as defined in Chapter 1 of this book, it is necessary to establish some way of connecting the electrical system to the earth. What is needed is a conductive (normally metallic) object that will adequately dissipate fault currents into the surrounding earth, although, as we will see, the earth is not an equipment grounding conductor. The actual physical connection to earth takes place at a grounding electrode.

The NEC clearly defines preferred grounding electrodes as well as other electrodes permitted when these preferred types are not available. Rules covering this subject are found in Part C of Article 250. These will be discussed first to provide a broader understanding on the subject of grounding as well as serve as a basis for NEC rules that cover system and equipment grounding.

SEC. 250-50 – GROUNDING ELECTRODE SYSTEM

To connect an electrical system or equipment that is required to be grounded to a grounding electrode, there must be some way of connecting the two together. An unspliced conductor called a grounding electrode conductor, must connect the objects to be grounded to the grounding electrode(s). The unspliced grounding electrode conductor is permitted by this section to run to, and be connected to, any of the grounding electrodes listed in Sec. 250-50(a) - (d) and 250-52(b) - (d). The grounding electrode conductor must be unspliced unless it is spliced by means of an irreversible compression-type connector listed for the purpose, or by the use of exothermic welding to join the sections.

The title of Part C refers to a grounding electrode *system*. While the system can consist of a single driven ground rod or similar single device, most often the grounding electrode system consists of a larger interconnected mass of metallic items. When any of the grounding electrodes listed in Sec. 250-50(a) - (d) and Sec. 250-52(b) - (d) are available at a location, they *must* be bonded together to form the grounding electrode system. See **Fig. 2.1** for a graphic illustration of these connections.

Note that the word "available" is used in the previous sentence instead of the word "present". Formal interpretations have clarified that steel reinforcing bars (a form of concrete-encased electrode) need not be exposed after a concrete foundation has been poured to bond the bars to the electrode system.

The topic of bonding is discussed in depth in Chapter 3 of this book. In this part of the code, the following references are made to bonding techniques that must used to form a grounding electrode system.

• Sec. 250-64 for installation
• Sec. 250-66 for sizing

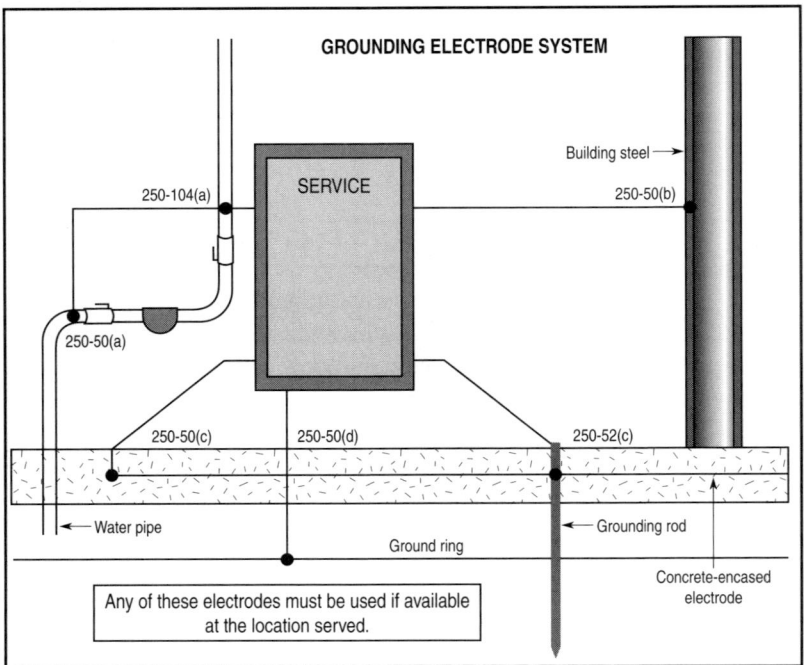

GROUNDING ELECTRODE SYSTEM

Building steel →

250-104(a)   SERVICE   250-50(b)

250-50(a)

250-50(c)   250-50(d)   250-52(c)

← Water pipe   ← Grounding rod

Ground ring

Concrete-encased electrode

Any of these electrodes must be used if available at the location served.

Fig. 2.1 Any of these electrodes must be used if available at the location served.

- Sec. 250-68 for connections

Permission by NEC to use a cold water pipe as part of a grounding electrode system has undergone changes over time because of the increased use of plastic piping by water-supply utilities. This is often done after the original installation, and without appropriate understanding of or appreciation for the hazards that ensue when a key part of a grounding electrode system is removed without warning to the owner and to local electrical authorities.

There is another key restriction on the use of water pipe as an electrode (see **Fig. 2.2**). No point on a water pipe beyond 5 ft from the building entrance is eligible for connection, either to the principal grounding electrode conductor, or to interconnect other local grounding electrodes that are part of the grounding electrode system.

The only exception to this occurs in industrial and commercial occupancies. There, if the entire length of the water pipe proposed for use as a conductor is exposed, and if there is qualified maintenance and supervision, then remote connections are allowed. Be aware that the literal text refers to the "entire length" being exposed and you may need that

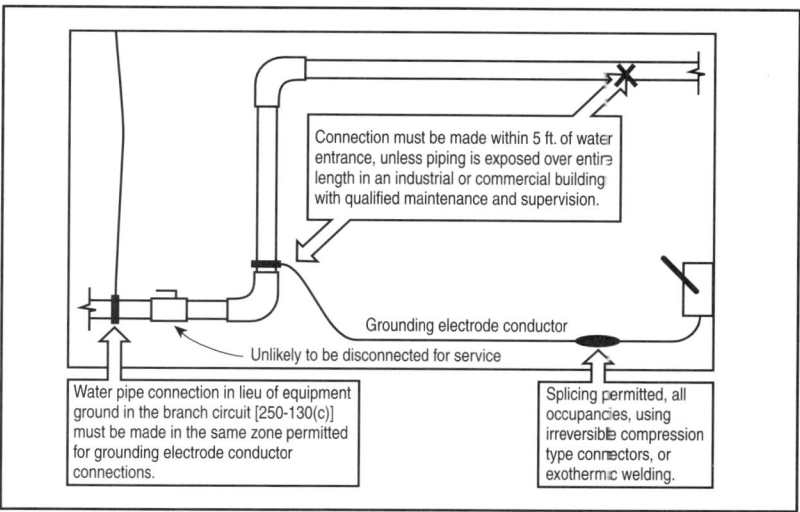

Fig. 2.2 The restriction against remote connections to water pipes as electrodes affects many other Code rules.

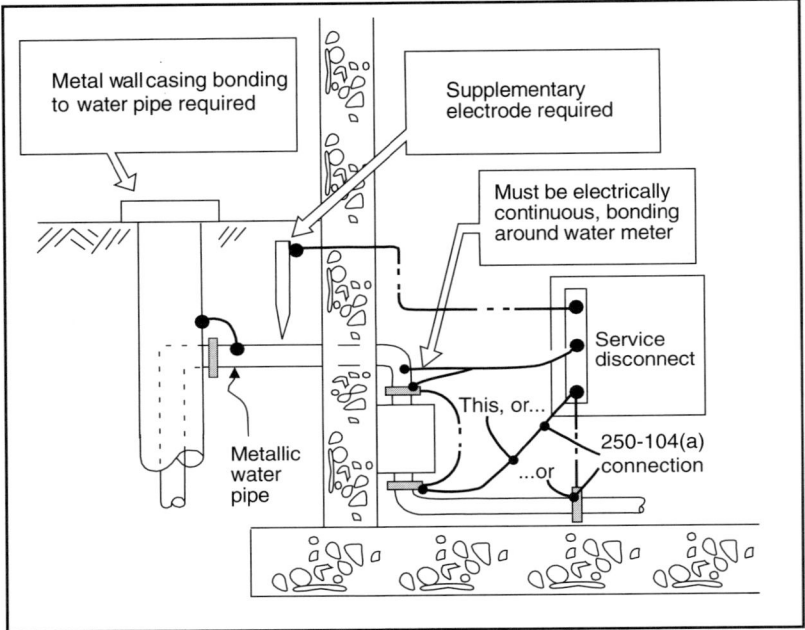

Fig 2.3 A metal underground water pipe with at least 10 ft in direct contact with the earth is an acceptable grounding electrode.

conductor to pass through firestopped partitions. The panel intends that such passages shouldn't kill the exception, but they also rejected a proposal to allow clarify it, believing it to be obvious. Make sure your local authority agrees.

The Code is responding to documented cases where grounding connections were disrupted by other trades and ignored. Note that "exposed" is not the same as "visible". Under the definitions in Art. 100, a water pipe run above a suspended ceiling, although not normally visible, would still meet the definition of "exposed."

(a) Metal underground water pipe. In the past, a cold water pipe entering a facility from the outside and connected to a metallic water main that is buried and in direct contact with the earth was accepted as being suitable to serve as the grounding electrode. As long as such a water piping system remains in place, it is one of the best electrodes available. Many water supply agencies, however, are replacing these systems with nonmetallic systems; and when this happens, the ground connection is lost. Since it is possible to lose this grounding connection during the life of a structure, the NEC now requires that the water pipe must be supplemented by some other electrode. It now requires that if a metal underground water pipe is used as a grounding electrode, as shown in **Fig. 2.3**, it must be:

- in direct contact with the earth for 10 ft. (3.05 m) or more;
- electrically continuous to the points of connection of the grounding electrode conductor and bonding conductors (or made electrically continuous by bonding around insulating joints or sections of insulating pipe); and
- supplemented by another electrode of a type specified in Secs. 250-50 or 250-52. If the supplementary electrode is a rod, pipe, or plate, then it must meet the resistance restrictions in Sec. 250-56 (see **Fig. 2.4**).

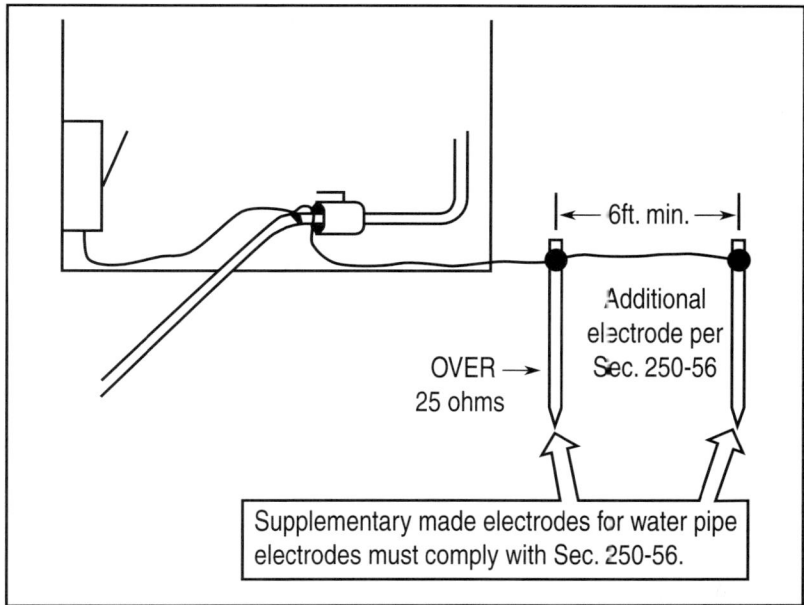

Fig. 2.4 These ground rods must be installed so that they would constitute a qualifying grounding electrode system with the water pipe removed.

Be careful. This has the effect of requiring an additional rod or pipe electrode in those frequent instances in which the ground resistance of the first electrode exceeds 25 ohms. You can't have your cake and eat it too. Some are suggesting that since Sec. 250-56 allows for water pipes to augment a deficient made electrode, the water pipe plus the single electrode are enough. There are two problems with this. The first is that that is saying that the electrode that needs to be supplemented is capable of being a supplement, which is circular reasoning at best.

The second, far more important reason is that that argument overlooks the reason for the rule in the first place. As long as the water pipe is in place, the ground rod is a mouse helping to carry the burden of an elephant. Its only function is to serve as the principal electrode if the metal water pipe is ever replaced with plastic. In that sense, a better term for it would be a reserve electrode. The intent of this change is absolutely clear: the made electrode system installed under this section must fully qualify as an electrode system with the metal water pipe removed.

The grounding path or the bonding connection to interior piping must not rely on water meters for continuity. In other words, a bonding jumper must be used around a water meter. The same rule applies to water filtration equipment and the like.

A supplemental electrode is permitted to be bonded to the grounding electrode conductor, the grounded service-entrance conductor, the grounded service raceway, or any grounded service enclosure.

A parenthetical note in Sec 250-50(a) mentions that a metal well-casing that is effectively bonded to the water-pipe electrode can be considered to be a part of the 10 ft. length required. This can be misleading. This is not saying that a well casing is a water-pipe electrode. The requirement calls for a bonding jumper from an otherwise qualified water pipe to an enclosing well casing should one be in use. This acts to prevent a "choke" effect of a single conductor (the water pipe) entering a metal enclosure (the well casing). In effect, the well casing ends up bonded to the grounding electrode at both ends, to the extent practicable. This is a fundamental safety concept associated with running grounding conductors, discussed further in Chapter 8 of this book.

(b) Metal Frame of a building. The metal frame of a building, if effectively grounded, is also a preferred grounding electrode and must be

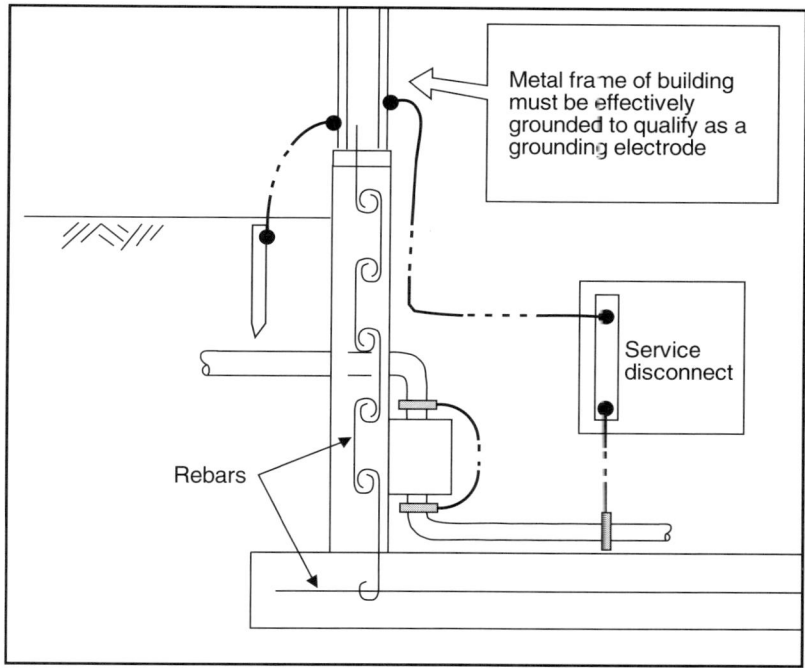

Fig. 2.5 Effectively grounded metal columns that are part of a building's framing are often used as grounding electrodes.

bonded to any of the other electrodes that are present at the site. It can also act as the supplemental electrode to a cold water pipe.

The definition in Art. 100 explains that "effectively grounded" means that the steel must have an intentional low-impedance connection to ground with sufficient current-carrying capacity to prevent the build-up of voltages that could harmful to persons or to equipment. In addition to the grounding method shown in **Fig. 2.5**, effectively grounding the building steel can be accomplished through building column connections to conductive elements in the building footings.

(c) Concrete-encased electrode. A preferred electrode can also be an electrode encased within at lease 2 inches (50.8 mm) of concrete. This type of electrode must be located within and near the bottom of a concrete foundation or footing that is in direct contact with the earth. This usually puts the electrode near the outside of the building where rain keeps the soil moist.

This type of electrode is often referred to as a Ufer system. It can consist of at least 20 ft. (6.1 m) of one or more steel reinforcing bars or rods that are not less than ½-in. (12.7 mm) diameter (see **Fig 2.6A**). The bars or rods can be bare, zinc coated, or coated with another electrically conductive coating.

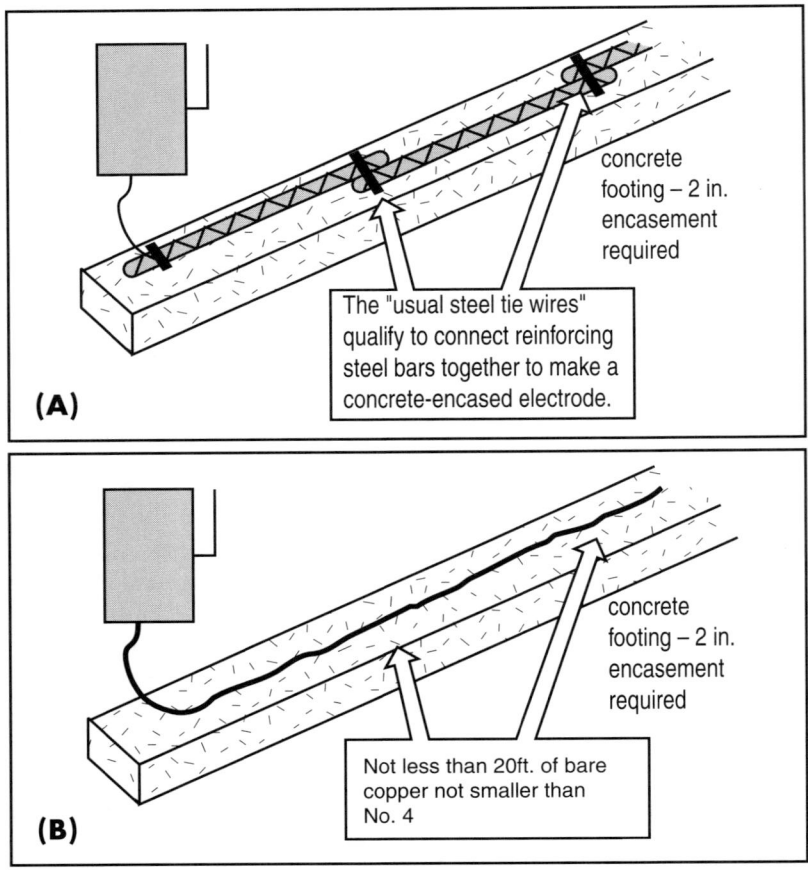

(A)

concrete footing – 2 in. encasement required

The "usual steel tie wires" qualify to connect reinforcing steel bars together to make a concrete-encased electrode.

(B)

concrete footing – 2 in. encasement required

Not less than 20ft. of bare copper not smaller than No. 4

Fig. 2.6 Rebars within a footing **(A)**, or bare copper conductor **(B)** encased in concrete both make effective grounding electrodes.

An alternative to the use of reinforcing bars or rods in the footing is the use of at least 20 ft (6.1 m) of bare copper conductor, not smaller than No. 4 (see **Fig 2.6B**). If the footing has not been poured (thus is still available), this is an outstanding electrode. Even if steel reinforcing is not used, the electrode can be installed more easily and at lower cost than typical made electrodes. All that is needed is a suitable length of bare No. 4 copper wire to lay into the footing. For example, 20 ft. of cable could be in the footing; the remaining length could exit the footing and run directly to the service equipment. This would make the ground reliable (no splicing) and avoids the necessity of a separate connection to a grounding electrode conductor where it emerges from the footing.

It isn't often that we get a far better result for spending less money on material and investing less time. A concrete encased electrode may allow for just that. You don't have the expense on installation time involved with ground rods and clamps. Not only that, the result is usually

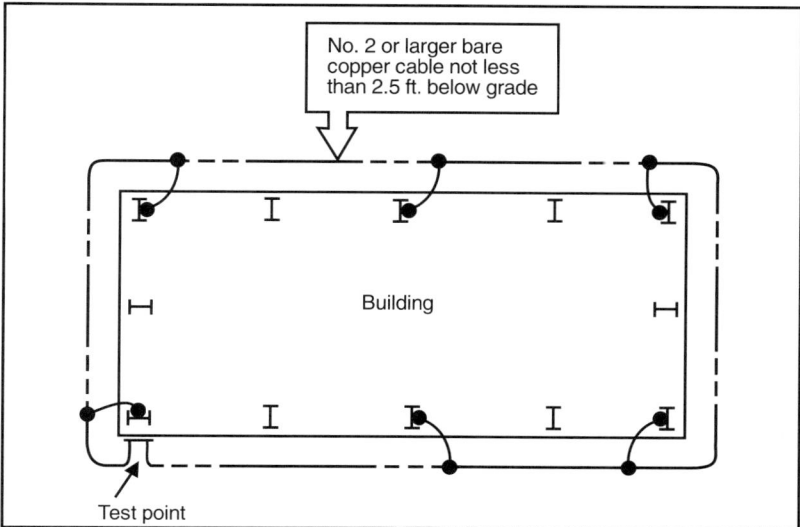

Fig. 2.7 A ground ring consisting of a No. 2 or larger bare copper cable buried at least 2½ ft in the earth is considered to be a good grounding electrode. Where a metal-frame building is used, usually every other column is connected to the ring. A ground ring is equally effective as a point for connecting lightning protection down conductors used for concrete or masonry structures.

on the order of ten times better in terms of lower ground resistance.

(d) Ground ring. Another option is for the supplemental electrode to be 20 ft. (6.1 m) minimum of bare copper cable, not smaller than No. 2, that completely encircles the building. The cable must be buried not less than 2.5 ft (762 mm) below grade as shown in **Fig 2.7.**

SEC. 250-52–MADE AND OTHER ELECTRODES

If none of the grounding electrodes specified in Sec. 250-50 are available, then the code allows other types of electrodes to be used for grounding purposes. These made electrodes must be embedded in the ground below the permanent moisture level, if possible, and they must be free from paint or other nonconductive coatings.

Where more than one electrode is used, an electrode of one grounding system must not be less than 6 ft. (1.83m) from any other electrode of another grounding system. Sec. 250-52 explains that two or more electrodes that are effectively bonded together are considered to be a single grounding electrode system in interpreting this requirement. FPN No. 2 of 250-106 clarifies that if there is a grounding electrode system around a building, a separate system intended to meet the requirements for another building, or one used for lightning protection, etc., it must be kept at least 6 ft. from the first; or the two must be bonded together.

The following subsections of Sec. 250-52 list the types of other electrodes permitted to serve as a grounding electrode in a grounding electrode system.

b) Other local metal underground systems or structures such as an uncoated underground tank or piping system can be used as a grounding electrode. A metal well casing falls into this category and is an excellent electrode. Note that unlike other made electrodes, these electrodes require full sized grounding electrode conductors. The provisions in Sec. 250-66(a) for a No. 6 to be the largest required size for connecting to made electrodes does not apply here.

c) Rod and pipe electrodes are among the most common types of made electrodes. Rods, pipe, or conduit, in lengths of 8 feet or more, can be driven into the ground to serve as a grounding electrode. The following restrictions apply (see **Fig. 2.8**):

- If the electrode is made up of pipe or conduit, it must be at least ¾-in.

trade size in diameter, and if of iron or steel, it must be galvanized or otherwise metal-coated to resist corrosion.

• Rods of iron or steel must be at least 5/8-in (15.87 mm) diameter. Stainless steel rods, nonferrous rods, or their equivalent, are permitted to be less than 5/8-in., but must have a diameter of not less than ½-in. (12.7mm) and must be listed.

• Rod or pipe electrodes must be in contact with the soil for a length of at least 8 ft. (2.44 m).

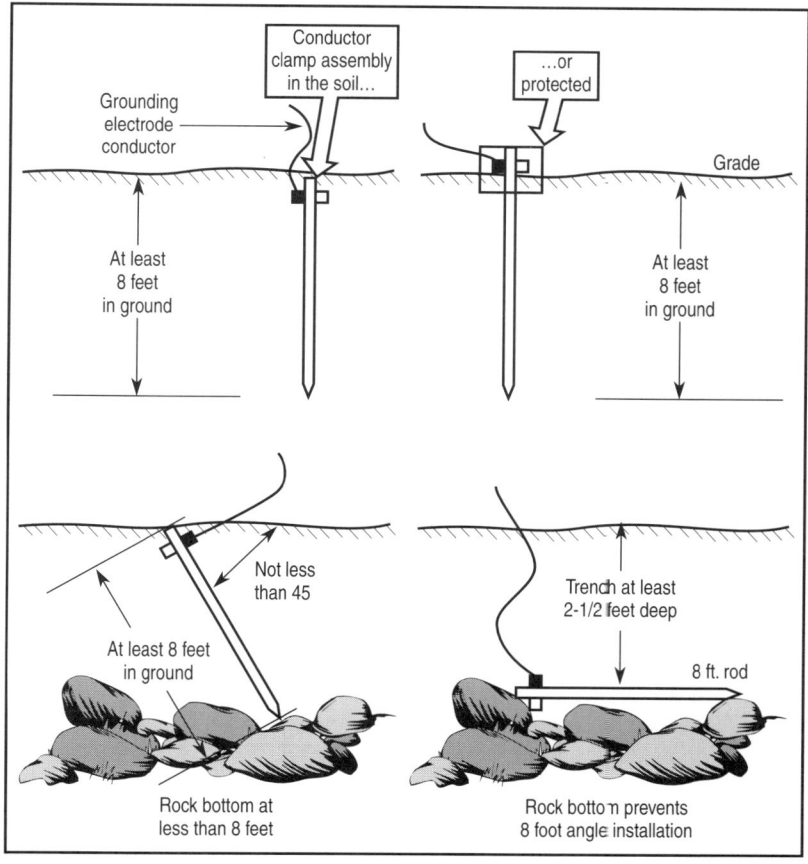

Fig. 2.8 Rod and pipe electrodes must meet the requirements of Sec. 250-52(c)(3).

• If rock prevents the rod or pipe from being driven to a depth of at least 8 feet, then it should be driven in at an angle not exceeding 45° from the vertical.

• An alternative to driving the rod or pipe at an angle is to bury it in a trench that is at least 2½-ft. (762 mm) deep.

• The upper end of a driven electrode must be flush with, or below ground level, unless the above-ground end and the conductors attached to it are protected against physical damage by being located in a place where they are not likely to be damaged, or the conductors must be provided with a protective covering.

Note that the rule requiring 8 ft. burial in soil will put the end of an 8-ft. rod below grade. Only longer rods could have their ends exposed.

d) Plate electrodes must expose not less than 2 sq. ft. (0.186 sq. m) of surface to the soil. They must be at least ¼"-in. (6.35mm) thick if of iron or steel, or not less than 0.06-in (1.52mm) thick if of non-ferrous metal.

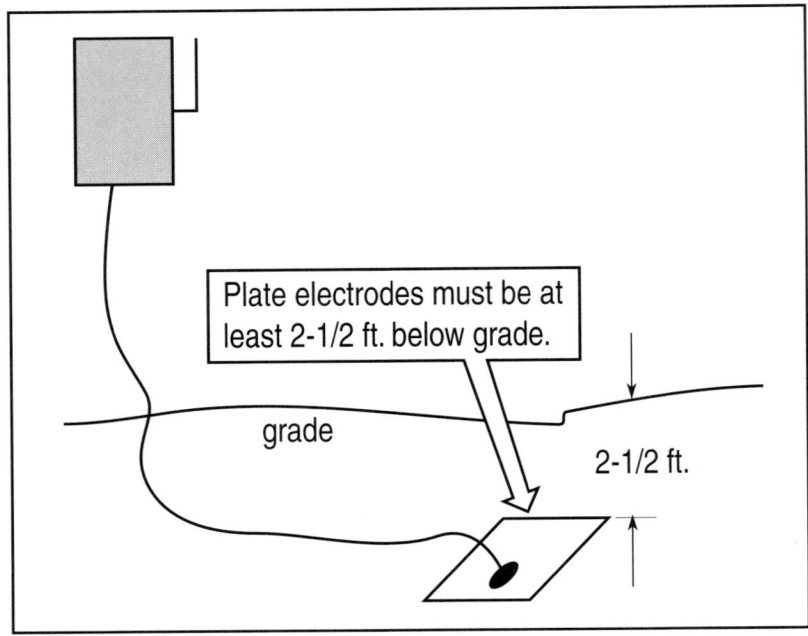

Plate electrodes must be at least 2-1/2 ft. below grade.

grade

2-1/2 ft.

Fig. 2.9 Plate electrodes are also used as made electrodes.

As shown in **Fig. 2.9**, they must be buried at least 2½-ft below grade.

The following subsections of Sec. 250-52 list the types of made or other electrodes that are prohibited from being used as grounding electrodes:

a) Metal underground gas piping systems

e) Aluminum electrodes

SEC. 250-56 – RESISTANCE OF MADE ELECTRODES

A single electrode consisting of a rod, pipe, or plate that does not have a resistance to ground of 25 ohms or less must be augmented by one additional electrode of any of the types specified in Sections 250-50 or 250-52 (change these section numbers). If after installing the second electrode, the combined resistance still exceeds 25 ohms, it is not mandated by the Code that additional electrodes be added. (It must be noted that frequently project specifications require that electrodes be added until the 25 ohm level is reached).

If this additional required electrode is a parallel rod, pipe, or plate, it must be located at least 6 ft. (1.83m) away from the first. An FPN states that when paralleling rods longer than 8 ft. (2.44 m), the efficiency of the combined electrodes is improved by spacing them further than 6 ft. apart.

SEC. 250-58 – COMMON GROUNDING ELECTRODE

Where an AC system is connected to a grounding electrode in or at a building as specified in Secs. 250-24 and 250-32, the same electrode must be used to ground conductor enclosures and equipment in or on that building. This requirement is discussed further in Chapter 5 of this book.

Where separate services supply a building and are required to be connected to a grounding electrode, the same grounding electrode must be used. Two or more grounding electrodes that are effectively bonded together are considered to be a single grounding electrode system in this sense.

This separate service requirement is a variation of the situation described in Sec. 250-24(a)(3). There, the two services were tied together by a secondary tie and a single electrode was *permitted* to be used. Here,

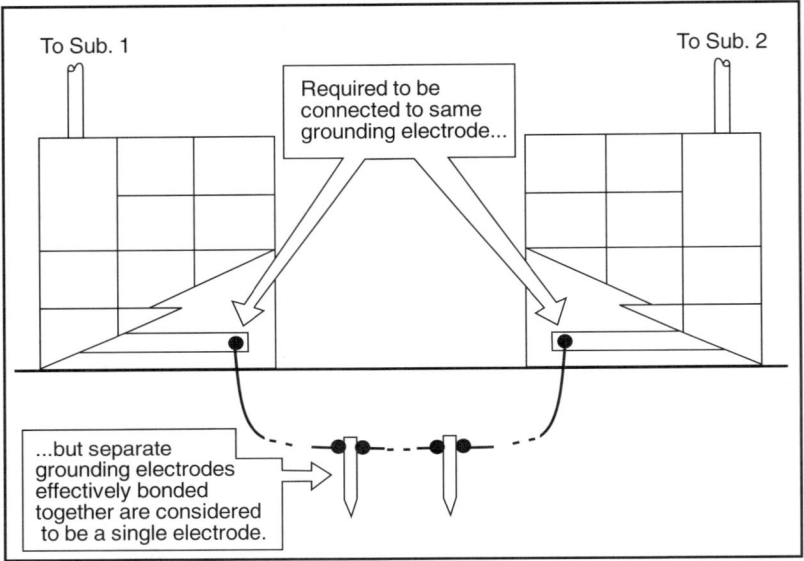

To Sub. 1

To Sub. 2

Required to be
connected to same
grounding electrode...

...but separate
grounding electrodes
effectively bonded
together are considered
to be a single electrode.

Fig. 2.10 Two or more effectively bonded grounding electrodes are considered to be a single grounding electrode. Two separate connections to a grounding ring also meet the requirement of Sec. 250-58.

as shown in **Fig 2.10**, the separate services are not tied together and are *required* to be connected to a common grounding electrode. If two separate electrodes are used, then they must be effectively bonded to one another. This applies even on services of different voltage systems, or at opposite ends of a large building, as covered in Sec. 230-2(b)(2) and Sec. 230-2(d).

SEC. 250-60–USE OF AIR TERMINALS FOR LIGHTNING PROTECTION

Lightning protection down conductors, driven pipes or rods, or other made electrodes used for grounding lightning protection systems must not be used in place of the made grounding electrodes required by Sec. 250-52 for grounding wiring systems and equipment as shown in **Fig. 2.11**.

This provision, however, does not prohibit the bonding together of grounding electrodes of different systems. As FPN No. 2 to this section

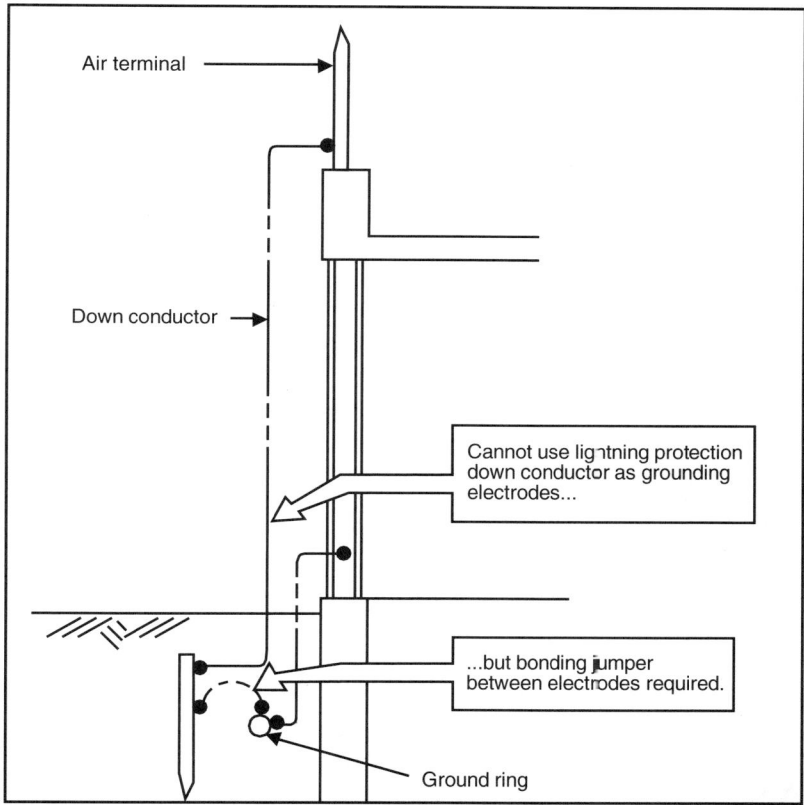

Air terminal

Down conductor

Cannot use lightning protection down conductor as grounding electrodes...

...but bonding jumper between electrodes required.

Ground ring

Fig. 2.11 Lightning protection conductors and equipment and electrical system grounding must not use the same made electrodes. They, however, must be bonded together.

points out, bonding together of all separate grounding electrodes will limit potential differences between them and between their associated wiring systems. Furthermore, Sec. 250-106, which we cover in the next chapter, makes this bonding mandatory.

SEC. 250-62 – MATERIAL

This section covers the material from which the grounding conductors can be made.

Grounding electrode conductors can be made of copper, aluminum, or copper-clad aluminum. The material selected must be resistant to any corrosive condition existing at the installation or be suitably protected against corrosion. The conductor can be solid or stranded, insulated, covered, or bare. There is no required or ezpressly prohibited color for these conductors, although Sec. 200-7 generally reserves the color white for grounded conductors. These are not grounded, because they aren't current carrying. The rules in Sec. 250-119 would not, however, preclude using the color green.

SEC. 250-64(a)–INSTALLATION

The NEC prohibits aluminum or copper-clad aluminum conductors, if used outside, from coming within 18 in. of the soil. For indoor work, aluminum (or copper-clad aluminum) can't run in contact with soil or masonry or other corrosive conditions.

(a) **Grounding electrode conductors** and their enclosures must be securely fastened to the surface on which they are carried. In addition:

- A No. 4 or larger copper or aluminum conductor must be protected if exposed to severe physical damage;
- A No. 6 grounding conductor not subject to physical damage is permitted to run along the surface of the building construction without

One of the types of connectors that can be used to tap on to a grounding electrode conductor.

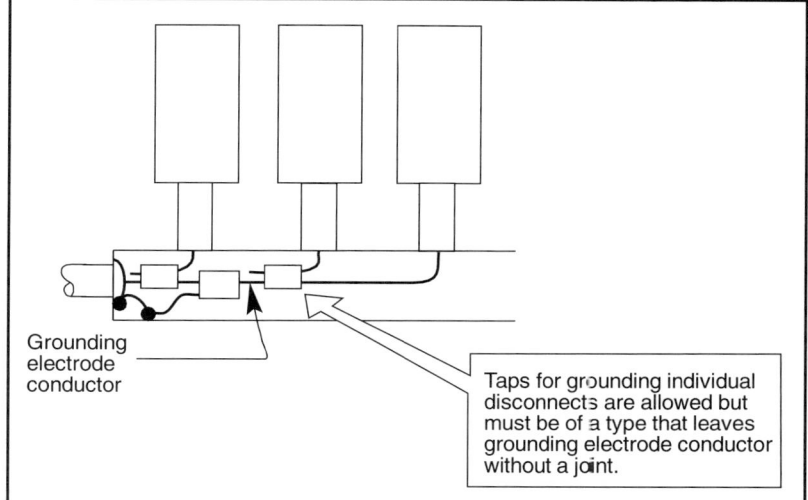

Grounding electrode conductor

Taps for grounding individual disconnects are allowed but must be of a type that leaves grounding electrode conductor without a joint.

Fig. 2.12 The general rule is that the grounding electrode conductor must be run in one continuous length from service disconnect to grounding electrode. Taps must be made without splicing.

metal covering or protection where it is securely fastened to the construction; otherwise, it must be in rigid metal conduit, IMC, rigid nonmetallic conduit, EMT, or cable armor; and

• Grounding conductors smaller than No. 6 must be in rigid metal conduit, IMC, rigid nonmetallic conduit, EMT, or cable armor.

A grounding electrode conductor should be installed in one continuous length without a splice or joint. There are, however, several exceptions to this continuous-length requirement. An exception permits splices in busbars. In addition to busbars, that are inherently jointed, the Code [Sec. 250-64(b)] also allows for exothermic welding and for irreversible crimp connectors listed for the purpose.

(b) Grounding electrode conductor taps are required in cases where the service disconnect set consists of two to six separate units, as permitted by Sec. 230-40, Exception No. 2. Sec 250-64(c) permits taps to be made to the grounding electrode conductor for grounding individual disconnecting devices. The tap conductors must be connected to the grounding electrode conductor in such a manner that the grounding electrode conductor remains without a splice or joint. See **Fig. 2.12** for an illustration of properly made taps.

The sizing of the grounding electrode conductor and tap conductors is covered in Chapter 11 of this book.

Sec 250-64(a) – Grounding electrode conductors and their enclosures must be securely fastened to the surface on which they are carried. In addition:

• A No. 4 copper or aluminum conductor must be protected if exposed to severe physical damage;

• A No. 6 grounding conductor not subject to physical damage is permitted to run along the surface of the building construction without metal covering or protection where it is securely fastened to the construction; otherwise, it must be in rigid metal conduit, IMC, rigid non-metallic conduit, EMT, or cable armor; and

• Grounding conductors smaller than No. 6 must be in rigid metal conduit, IMC, rigid nonmetallic conduit, EMT, or cable armor.

Sec. 250-64(d) Enclosures for grounding electrode conductors are required to be electrically continuous from the point of attachment to cabinets or equipment to the grounding electrode, and must be securely fastened to the ground clamp or fitting. Metal enclosures that are not physically continuous as required must be made electrically continuous by bonding *each end* to the grounding conductor.

This is absolutely critical in ac systems, because the current always tries to flow over the outermost surface due to skin effect, and in this case that means over the enclosing raceway. If both ends aren't bonded, then you force the current over the interior conductor, but at great cost in

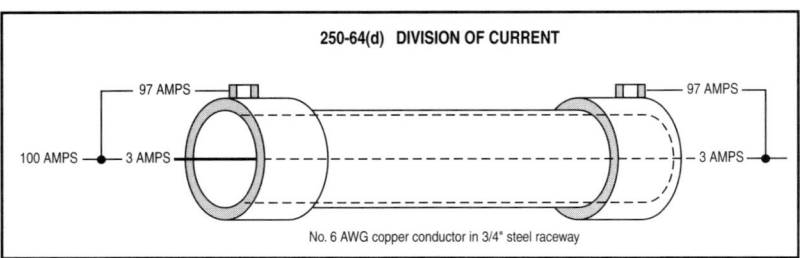

Fig. 2.13 AC grounding currents prefer to flow over the enclosing raceway. If the bonding connections are broken at either end, the impedance roughly doubles.

impedance. **Fig. 2.13** shows an actual test set-up to verify this.

Where a raceway is used as protection for a grounding conductor, the installation must comply with the requirements of the NEC article that applies to the raceway.

SEC. 250-68 – GROUNDING ELECTRODE CONDUCTOR CONNECTIONS

The mechanics by which grounding continuity is to be maintained is of great importance. Safety of persons working on or near electrical equipment depends upon the continued adequacy of the low impedance path over which fault current will flow. While it may test "good" immediately following installation, the grounding system may deteriorate over time, thus endangering those it was intended to protect. If the rules contained in this section and

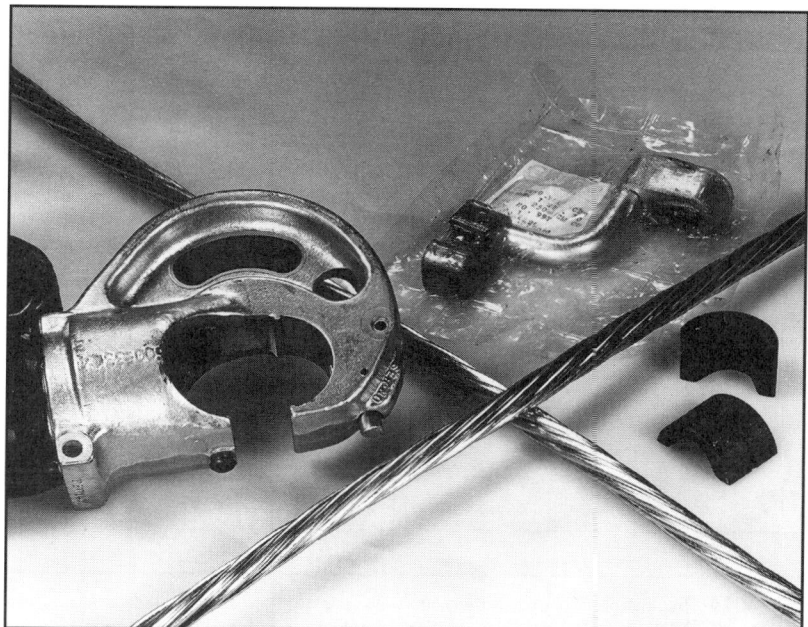

Pressure-type connectors insure that the current-carrying capacity of the connections exceed that of the conductors.

the immediately following sections of the Code are adhered to, the possibility of grounding system discontinuity will be greatly reduced.

The connection of a grounding electrode conductor to a grounding electrode must be accessible and made in a manner that will assure a permanent and effective ground.

An exception is made to this "assessibility" rule when the grounding electrode is concrete-encased, driven, or buried. In these cases, the connections are not required to be accessible.

An ongoing problem with grounding reliability has to do with interference in grounding paths by nonelectrical personnel. To minimize this, be sure that bonding jumpers around equipment that my be removed for service are long enough to allow the other trade to work. **Fig. 2.14** shows a good example.

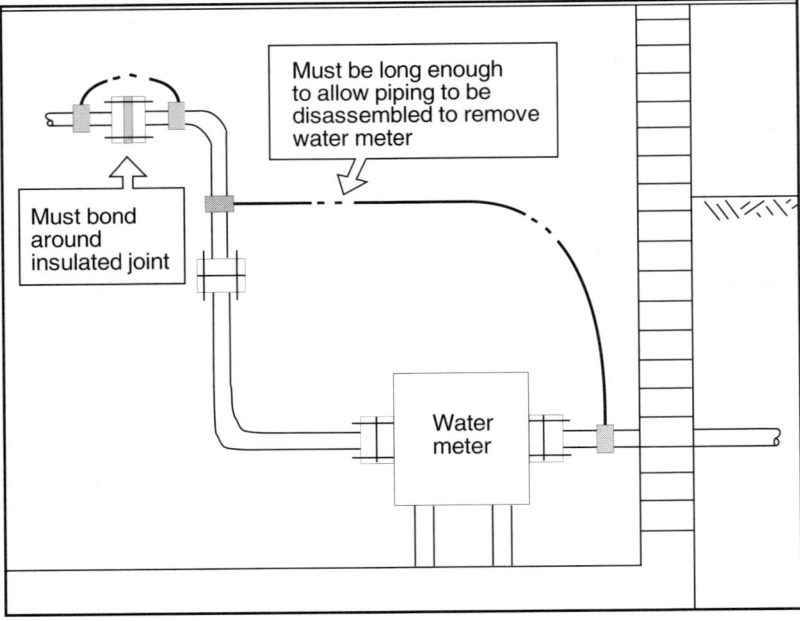

Fig. 2.14 The installation of bonding jumpers is essential to the maintenance of an effective grounding path. Special care must be taken when bonding around a piece of removable equipment to assure that when the item is removed, the continuity of the grounding path will not be interrupted.

SEC. 250-70-GROUNDING ELECTRODE CONNECTIONS TO ELECTRODES

The grounding electrode conductor must be connected to the grounding electrode by:

- Exothermic welding;
- Listed lugs;
- Listed pressure connectors;
- Listed clamps; or
- Other listed means.

Connections depending on solder must not be used. This section includes four specific methods as follows, and general requirements.

(1) Bolted clamp. A listed bolted clamp of cast bronze or brass, or plain or malleable iron.

(2) Pipe fitting, pipe plug, etc. A pipe fitting, pipe plug, or other approved device screwed into a pipe or pipe fitting.

(3) Sheet metal strap-type ground clamp. A listed sheet metal clamp of this type having a rigid metal base that seats on the electrode and

Sheet metal ground clamp. Those of the strap type (not shown) are for indoor telecommunications purposes only.

The general rule is that only one connection to a grounding electrode can be made with a single clamp. More connections, however, can be made if the clamp is listed for multiple connections.

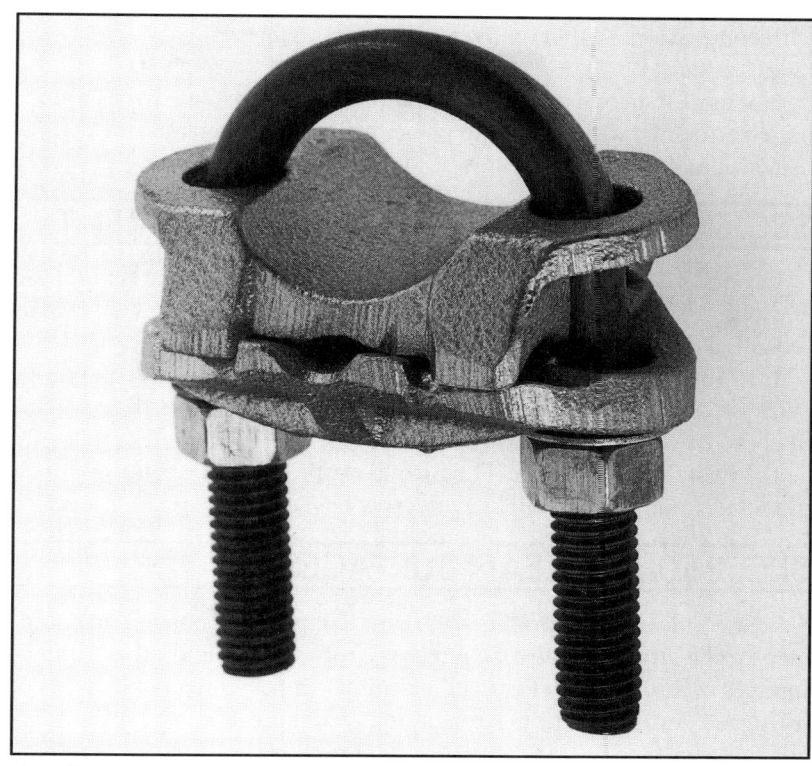

Listed connectors should be used for connecting to grounding electrodes. Only connectors marked "copper water tubing" or equivalent can be used with sweated copper water tubing systems.

having a strap of such material and dimensions that it is not likely to stretch during or after installation. *This method is only allowed for indoor telecommunications purposes,* and not for power systems.

(4) Other means. An equally substantial approved means.

Ground clamps must be listed for the materials of the grounding electrode and the grounding electrode conductor. Where they are used to connect to pipe, rod, or other buried electrodes, they must also be listed for direct soil burial. You can tell because suitable clamps are marked "direct burial" or "DB" if abbreviated. Such clamps will be either copper, brass, or bronze, or they will be made of stainless steel. If you see a clamp with conventional galvanized screws, for example, it probably isn't listed for direct burial.

Not more than one conductor is allowed to be connected to the grounding electrode by a single clamp or fitting unless the clamp or fitting is listed for multiple conductors.

SEC. 250-10–PROTECTION OF ATTACHMENT

Ground clamps or other fittings must be approved for the general use without protection, or they must be protected from physical damage as indicated in the following subsections.

(a) Not likely to be damaged. The fittings can be installed without extra protection from physical damage in locations where they are not likely to be damaged.

(b) Protective covering. The fittings are protected by being enclosed in metal, wood, or equivalent protective coverings.

SEC. 250-12–CLEAN SURFACES

Nonconductive coating such as paint, lacquer, and enamel on equipment to be grounded must be removed from threads and other contact surfaces to assure good electrical continuity or be connected by types of fittings so designed as to make such removal unnecessary.

3

ARTICLE 250, PART E – BONDING

Bonding is extremely important to the overall scheme of effective grounding. Once this topic and its Code rules have been clarified, the remaining parts of Article 250 will be addressed .

Bonding is a distinct term and is not to be used in place of the word "grounding." It is, instead, a technique for assuring the electrical continuity of a low impedance grounding path. It is defined in Article 100 as the permanent joining together of all metallic parts of the wiring system – such as conduit, fittings, boxes, and enclosures – to form an electrically conductive path that assures electrical continuity and the capacity to conduct safely any current likely to be imposed. Most frequently, bonding is used to assure a complete ground fault current return path.

Other related definitions in Article 100 include:

• **Bonding jumper**, defined as being a reliable conductor to assure the required electrical conductivity between metal parts required to be electrically connected.

• **Bonding jumper, equipment**, is the connection between two or more portions of the equipment grounding conductor.

• **Bonding jumper, main**, is the connection between the grounded circuit conductor and the equipment grounding conductor at the service.

SEC. 250-90–GENERAL

As part of a general rule, the Code requires that bonding be provided where necessary to assure electrical continuity and the capacity to safely conduct any fault current likely to be imposed. This is basically a restatement of the definition of bonding given in Article 100.

SEC. 250-92–SERVICES

Nowhere is the need for a low-impedance path for ground fault currents more important than at the service entrance. It is here that the electrical system is connected to the grounding electrode system. From anywhere in the building's electrical system, fault current must be able to flow unimpeded back to the ground connection at the energy source (normally the neutral conductor of the transformer) via a predictable path rather than through a person touching a faulted piece of equipment.

The other issue is that conductors from utility sources are functionally

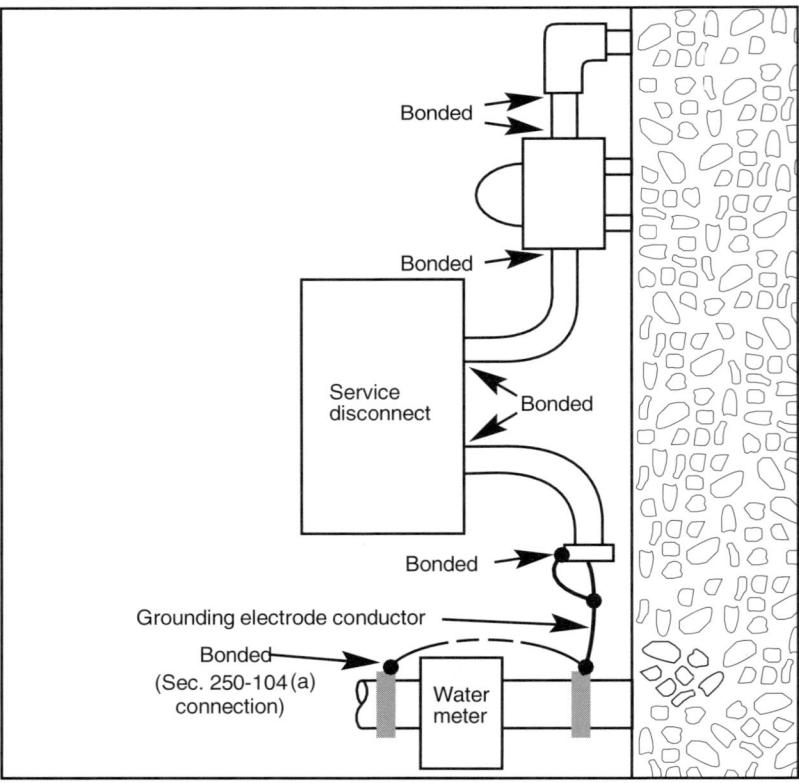

Bonded

Bonded

Service disconnect

Bonded

Bonded

Grounding electrode conductor

Bonded
(Sec. 250-104(a)
connection)

Water meter

Fig. 3.1 All components on the line side of the service equipment must be bonded together to form an electrically continuous system.

unprotected. In case of a service fault, a well bonded system maximizes current discharge to the grounding electrode and to the grounded service conductor. This, in turn, provides the best opportunity for such a fault to burn clear quickly, and with the lowest possible voltage to ground on exposed electrical enclosures.

The Code, therefore, spells out the requirements for bonding at service entrances in the following subsections.

(a) Bonding of services. All of the following noncurrent-carrying metal parts of service equipment are required to be effectively bonded together, as illustrated in **Fig. 3.1**:

• Raceway or support system(s) of service conductors such as conduit, cable tray, busway enclosure, cablebus framework, or other systems used for the purpose.

• Meter boxes, fittings, or other enclosures in the raceway or support system that contains the service conductors.

• Metallic raceway or armor used to protect a grounding electrode conductor from damage. Bonding is to be used at each end of the run, and at all intervening raceways, boxes, and enclosures between the service equipment and the grounding electrode.

Metal armor or the sheath of a service cable also must be bonded to the service equipment. This requirement, however, is modified by Sec. 250-84 to exclude cable that is part of a continuous underground metal-sheathed cable system in which the armor or sheath of the cable is metallically connected (bonded) to the underground system. In this case, the armor or sheath is not required to be bonded to the service equipment, and can be insulated from the interior conduit system as shown in **Fig. 3.2**. The purpose of this exclusion is to reduce the possibility of low-level current flow along the armor.

(b) Bonding to other systems. An additional bonding requirement is that the service must have an external provision to allow bonding and grounding conductors from other systems to be connected to the service grounding system. Several methods of making these provisions are listed, including:

• Exposed *nonflexible* metallic service raceways;

• An exposed grounding electrode conductor

• Approved means for the external connection of a copper or other

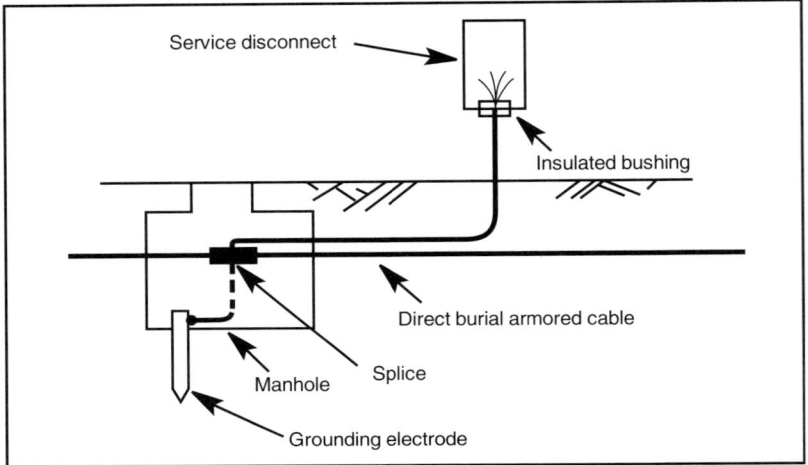

Fig. 3.2 Bonding of the metal sheath or armor of an underground service entrance cable to the service disconnect enclosure is not required when the armor is effectively grounded elsewhere.

corrosion-resistant bonding or grounding conductor to the service raceway or equipment. An FPN states that a No. 6 copper conductor with one end bonded to the service raceway or equipment, and with 6 inches (152 mm) or more of the other end made accessible on the outside surface is an example of this provision.

The most common types of intersystem bonding with grounding conductors connected to these external provisions are those of communications and CATV systems. Note that in the case of a separate building or structure fed from another (per Sec. 250-32), the disconnecting means for the separate building or structure is to be treated as if it were a service disconnecting means. The same principle holds for the disconnecting means for a mobile home, as covered in Sec. 550-23(a).

SEC. 250-94–METHOD OF BONDING SERVICE EQUIPMENT

There are many ways in which the required bonding of the service equipment and raceways can be achieved. The following subsections contain details on permitted methods.

(1) **Grounded service conductor.** An acceptable way to achieve bonding of service equipment is to make a connection to the grounded service conductor.

(2) **Threaded connections.** You can use threaded couplings, made up wrenchtight, as an acceptable method of bonding within the raceway system. Threaded bosses or hubs on enclosures can serve to bond the enclosure to entering raceways or metal-clad cable assemblies.

(3) **Threadless couplings and connectors** made up tight are also acceptable as the means of bonding metal raceways and metal-clad cables.

(4) **Other devices** that are approved for the purpose are permitted, such as bonding-type locknuts (to clean knockouts only); bonding wedges and bushings are also acceptable for bonding a raceway to an enclosure.

Bonding jumpers are defined as being reliable conductors that assure the required electrical conductivity between metal parts. They are permitted for bonding service equipment. Bonding jumpers must be used around concentric or eccentric knockouts that are punched or otherwise formed in a way that impairs the electrical connection to ground.

Standard locknuts or bushings are not permitted to serve as the bonding means for service equipment.

SEC. 250-96–BONDING OTHER ENCLOSURES

In addition to bonding requirements for items involved with services, the Code also requires all other noncurrent-carrying metallic components of a wiring system that are to act as equipment grounding conductor to be bonded to form a low-impedance path for fault currents.

Metal raceways, cable trays, cable armor, cable sheath, enclosures, frames, fittings, and other metal noncurrent-carrying parts that are to serve as grounding conductors must be effectively bonded where necessary to assure electrical continuity and the capacity to conduct safely any fault current likely to be imposed on them. Any nonconductive paint, enamel, or similar coating must be removed at threads, contact points, and contact surfaces or be connected by means of fittings that make such removal unnecessary.

Subsection (b) covers cases where electrical continuity is not desirable because of the need to reduce electrical noise on the grounding circuit. As shown in **Fig. 3.3**, an equipment enclosure is permitted to be

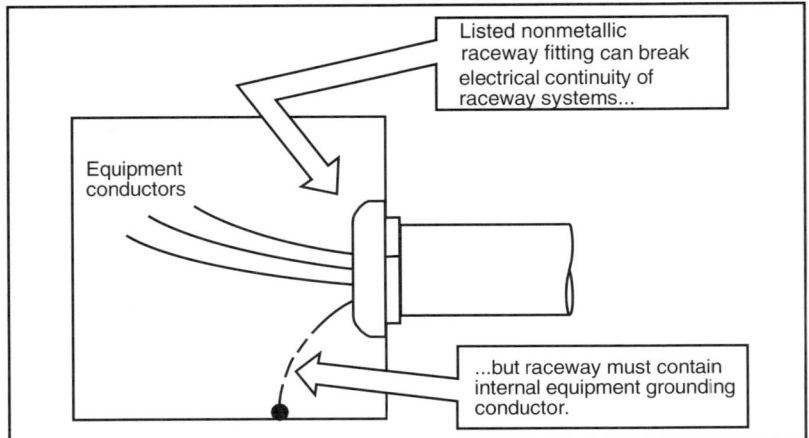

Fig. 3.3 Electrical continuity between raceway and enclosures can be broken to reduce electrical noise, but continuity must be maintained via an equipment grounding conductor run within the raceway. The raceway must be grounded in accordance with normal Code rules.

isolated from a raceway(s) by one or more listed nonmetallic raceway fittings located at the point of attachment of the raceway to the equipment enclosure. This permission, however, applies only if the raceway contains no wiring except that for the equipment in the enclosure. The listing requirement for the fitting should be interpreted to mean a listing as a raceway component, not specifically as an isolator.

Where a raceway and enclosure are isolated as allowed in these cases, the raceway must contain an internal insulated equipment grounding conductor to ground the equipment enclosure. In addition, an FPN serves as a reminder that the use of an internal insulated equipment grounding conductor does not eliminate the need for grounding the raceway system.

SEC. 250-97 – BONDING FOR OVER 250V

For circuits over 250V to ground, the electrical continuity of metal raceways and cables with metal sheaths that contain any conductor other than service conductors must be assured by one or more of the following:

- threaded connections;
- threadless couplings and connectors;

- bonding jumpers; or
- Other approved devices, such as bonding-type lock-nuts and bushings

The same rules as are shown in Sec. 250-94(b) - (e) for these items also apply here. There is, however, an exception made for installations where oversized, concentric or eccentric knockouts are not present. The following methods for assuring electrical continuity are permitted.

- For cables with metal sheaths: threadless couplings and connectors.
- For rigid metal conduit or IMC: two locknuts, one inside and one outside the box or cabinet.
- Fittings with shoulders that seat firmly against the box or cabinet (such as EMT connectors, flexible metal conduit connectors, and cable connectors) with one locknut on the inside of boxes or cabinets.
- Listed fittings for this purpose.

The same exception also allows omission of supplemental bonding around concentric or eccentric knockouts on systems over 250V to ground if the box or enclosure is listed for the use.

Small listed boxes with eccentric or concentric knockout patterns similar to that in the **Fig. 3.4** drawing will have been routinely evaluated for this use. This will be indicated in a marking on the smallest unit shipping carton for these boxes. Larger enclosures and pull and junction boxes (over 100 cu. in.) typically will not have been so listed, but if they are, there will also be a marking. Be sure to check with the manufacturer if there is any doubt. There has been some discussion about routinely evaluating all such knockouts and dropping the marking, but that will be in the future, if at all.

SEC. 250-98–BONDING LOOSELY JOINTED METAL RACEWAYS

Expansion joints and telescoping sections of raceways must be made electrically continuous by equipment bonding jumpers or other means.

SEC. 250-100–BONDING IN HAZARDOUS (CLASSIFIED) LOCATIONS

When a raceway and other enclosures of an electrical system are within an area classified as a hazardous location, the following techniques for

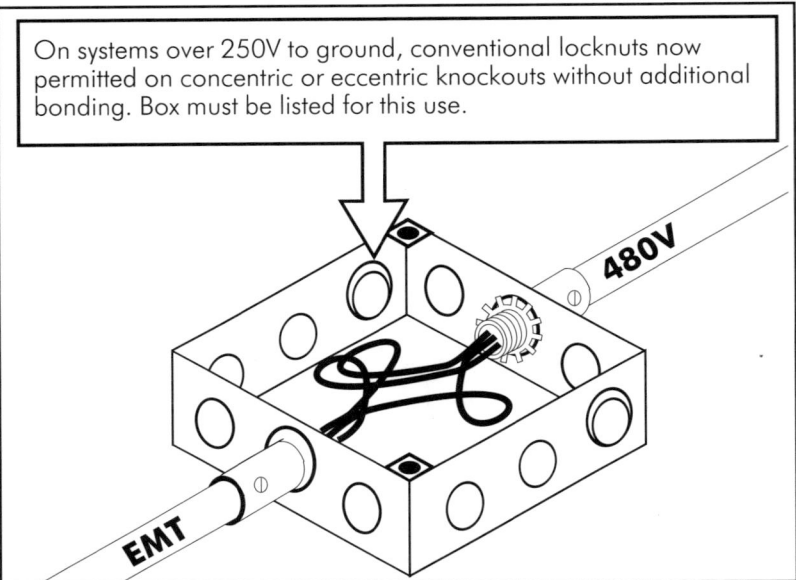

On systems over 250V to ground, conventional locknuts now permitted on concentric or eccentric knockouts without additional bonding. Box must be listed for this use.

480V

EMT

Fig. 3.4 Formerly all knockouts like the one shown required bonding jumpers over 250V to ground. This isn't necessarily true anymore.

bonding are allowed:
- threaded connections;
- threadless couplings and connectors;
- bonding jumpers; or
- other approved devices, such as bonding-type locknuts and bushings

Note that these methods are the same as permitted for bonding for over 250V in Sec. 250-97. Here, however, no exceptions apply. The definition and rules that apply to these bonding methods are the same as those given for these items in Sec. 250-94.

SEC. 250-102–EQUIPMENT BONDING JUMPERS

This set of rules specifies the material, sizing, and other details for bonding jumpers. They are listed in the following subsections.

(a) Material. Both the main and equipment bonding jumpers must be of copper or other corrosion-resistant material. The main bonding jumper must be a wire, bus, screw, or similar conductor.

(b) **Attachment**. Equipment bonding jumpers must be attached in the manner specified by the applicable provisions of Sec. 250-8. The attachment types permitted by that section includes:

- exothermic welding;
- listed pressure connectors;
- listed clamps, or
- other listed means.

Connection devices or fittings that depend solely on solder do not meet the requirement and cannot be used for this purpose. In addition, sheet metal screws can't be used to connect a grounding conductor to an enclosure.

Attachment of bonding jumpers to grounding electrodes must be in accordance with Sec. 250-70, which doesn't differ from Sec. 250-8 in this regard.

Sec. 250-102(c) and (d) detail how the size of bonding jumpers at the service are to be determined. As noted in the preface to this book, all rules for sizing grounding and bonding conductors, the tables that apply, and typical examples have been grouped within Chapter 11 of this book. Pulling all of this information into one chapter makes sizing easier to understand and locate.

(c) **Size** – equipment bonding jumper on supply side of service, and main bonding jumper.

(d) **Size** – equipment bonding jumper on load side of service

(e) **Installation** – equipment bonding jumper. The equipment bonding jumper is permitted to be installed inside or outside of a raceway or enclosure.

- Where installed on the outside, the length of the equipment bonding jumper must not exceed 6 ft. (1.83 m) and must be routed with the raceway or enclosure
- Where installed inside a raceway, the equipment bonding jumper must comply with the requirements of Secs. 250-119 and 250-148.

The equipment bonding jumper that is the subject of this subsection is most often required where a short length of flexible conduit that is not approved for grounding purposes is used to connect a piece of electrical equipment, as is shown in **Fig. 3.5**. It should run in a straight line, as shown, to reduce reactance.

Sec. 250-148 deals with equipment grounding conductor continuity and Sec. 250-119 deals with their identification. By reference to these two sections, the NEC requires that continuity of equipment of bonding jumpers run inside the raceway must not have to be interrupted to disconnect the equipment. The jumper insulation must be color-coded similar to an equipment grounding conductor.

SEC 250-104–BONDING OF PIPING SYSTEMS AND EXPOSED STRUCTURAL STEEL

Metallic piping other than that directly associated with an electrical system must also be bonded. This provides additional protection by reducing the possibility of shock by a person simultaneously contacting faulty electrical devices and a piping system grounded by other means. An FPN [in Sec. 250-104(c)] notes that bonding all piping and metal air

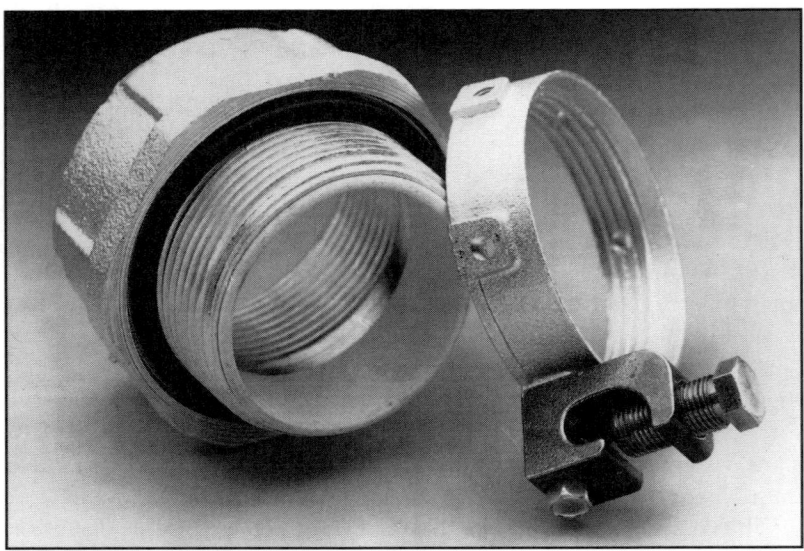

This bushing could be used for bonding to a box with conventional knockouts, but would need to be of the insulated type if used for conductors No. 4 or larger [Sec. 300-4(f)].

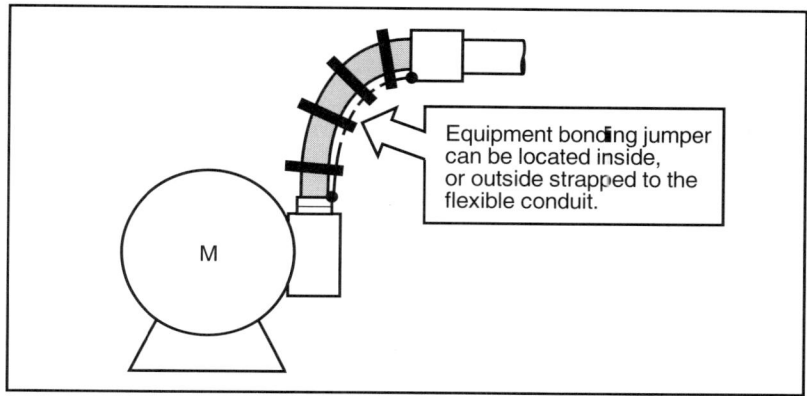

Equipment bonding jumper can be located inside, or outside strapped to the flexible conduit.

Fig. 3.5 When equipment grounding is accomplished via a raceway and its electrical continuity is interrupted, an equipment bonding jumper is required either inside or outside the raceway.

ducts within the premises will provide additional safety. Details are given in the following subsections.

(a) Metal water piping. The interior metal water piping system must be bonded to:

• the service equipment enclosure;
• the grounded conductor at the service;
• the grounding electrode conductor where of sufficient size; or
• to one or more of the grounding electrodes used.

The points of attachment of the bonding jumper must be accessible. This isn't always simple in cases where the distribution panel falls in a space with no plumbing. Often installers have run a bonding jumper to the nearest water pipe, which happened to be in a finished area, only to then have to install an unsightly access panel afterwards.

In addition, bonding jumpers used for this purpose must be installed in accordance with Secs. 250-64(a) and (d). Those subsections cover the installation of grounding electrode conductors. They require that the conductor be securely fastened to the surface on which it is carried and protected if subject to severe physical damage. The bonding jumper must also be electrically continuous. If run in a metal raceway, both ends of the raceway must be bonded to the enclosed conductor.

An exception to the need to bond the metal water piping is made in

Sec. 250-104(a)(2) for multiple occupancy buildings where the piping system for the individual occupancies is metallically isolated from all other occupancies by use of non-metallic water piping. In those cases, the interior metal water piping system for each occupancy is permitted to be bonded to the panelboard or switchboard enclosure (other than service equipment) supplying that occupancy.

Sec. 250-104(a)(3) covers a variation on this principle, where multiple buildings are fed from a common source, per Sec. 250-32. In these cases the local water piping system must be bonded to the building disconnecting means enclosure, or to the local grounding electrode, or to the supply equipment grounding conductor.

Sizing of the bonding jumpers for water piping systems is covered in Chapter 11 of this book.

Sec. 250-104(a)(4) covers bonding of local water piping systems to separately derived systems (Fig. 3.6). You must meet this bonding requirement regardless of the type of grounding electrode employed.

The requirement assures only minimal potential differences on nearby water piping systems, especially in view of Sec. 250-30(a)(3). That sec-

Metal piping systems that may become energized, such as this one under a raised floor of an information technology room, are required to be bonded to an appropriate part of the electrical system.

tion effectively requires in many cases that the actual grounding electrode connection for the separately derived system may be at a great distance from the local water piping. We cover separately derived system grounding in Chapter 5.

Sec. 250-104(b) requires aboveground portions of gas piping systems "upstream from the equipment shutoff valve" must be electrically continuous and bonded to the grounding electrode system (see **Fig. 3.7**).

This is extracted material from NFPA 54, the Fuel Gas Code. Although the NEC has always had a rule [Sec. 250-104(c)] to bond other metal piping, that rule refers to piping that is connected to electrical equipment and that may become energized thereby. Normally you automatically comply with that rule when you connect the equipment grounding conductor, because you can use the equipment grounding conductor

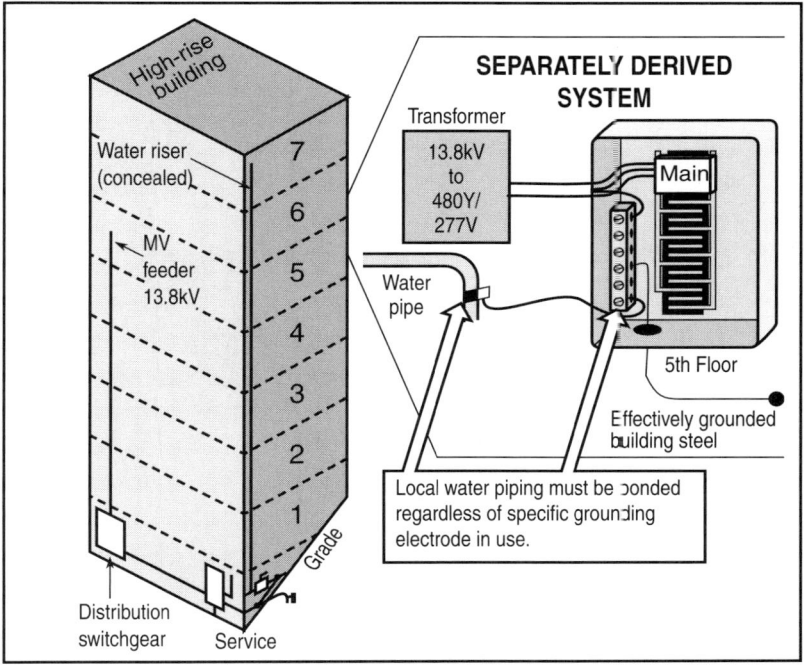

Fig. 3.6 Water piping systems in the same locality must be bonded to a separately derived system, at the same point where the GEC is connected.

of the supply circuit for bonding. This is different, because it applies even in cases where there isn't any electrical equipment connected to the gas piping system.

The rule doesn't say how big this conductor should be. In view of the fact that it applies to the gas piping system generally, the safest approach is to size in accordance with system bonding connections that also apply generally, such as to water piping systems. By that logic, size the conductor based on Table 250-66. You won't have to go back and do it again. Remember also, that a small size may be false economy. Sec. 250-64(a) as referenced in other subsections in Sec. 250-104 requires a raceway for sizes smaller than No. 6.

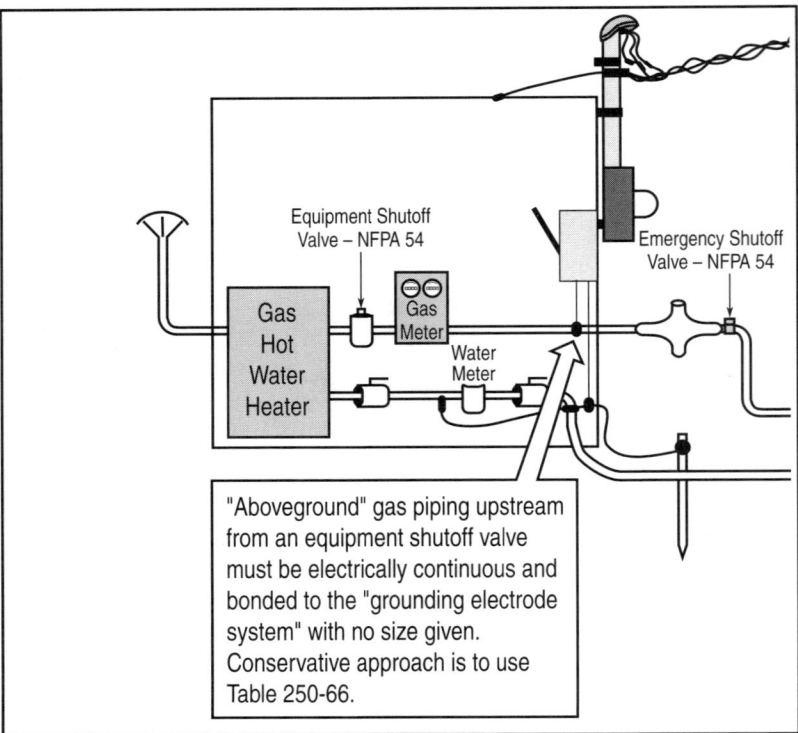

Fig. 3.7 The Code doesn't specify the size of this bonding conductor. A conservative approach that will work in all applications is to use Table 250-66.

Sec. 250-104(c) Other metal piping. Interior metal piping that may become energized must be bonded to:

- the service equipment enclosure;
- the grounded conductor at the service
- the grounding electrode conductor where of sufficient size; or
- to the one or more grounding electrodes installed.

The equipment grounding conductor for the circuit that may energize the piping is permitted to serve as the bonding means.

Sizing of the bonding jumper used for these metal piping systems is discussed in Chapter 11 of this book.

Sec. 250-104(d) Structural steel. Exposed interior structural steel that is interconnected to form a building frame, and that isn't "intentionally grounded" and which may become energized, must be bonded to the service grounding system (see **Fig. 3.8**). This requirement has to do with a building frame, not isolated sections of steel used for reinforcement or isolated girders for long framing spans, etc.

The framing must be bonded to the service equipment, or the grounded service conductor, or the grounding electrode conductor if of sufficient size, or to one or more of the grounding electrodes in the grounding electrode system for the building. As in the case of water pipe bonding, the bonding conductor is sized under Table 250-66 and installed in accordance with Sec. 250-64(a) and (d). The points of attachment must be accessible.

The substantiation for this requirement used the term "effectively" grounded, which is well defined as opposed to "intentionally" grounded which is what appears in the Code. Presumably the intent of the wording is, in fact, to pick exposed building steel frames that aren't "effectively" grounded. The connection is easily made, and since the framing is "interconnected" only one bonding connection should need to be made. There is some question as to how such a frame would be energized, since, like other piping systems covered in Sec. 250-104(c), a piece of electric equipment connected to it would have its own equipment grounding provisions.

SEC. 250-106–LIGHTNING PROTECTION SYSTEMS

The lightning protection system ground terminals must be bonded to the grounding electrode system.

Also, a FPN explains that metal raceways, enclosures, frames, and other noncurrent-carrying metal parts of electric equipment must be kept at least 6 ft. (1.83 m) in air or 3 ft. (0.92 m) through dense material such as concrete, brick, or wood away from lightning rod conductors. If this requirement cannot be met, then the option is to bond the items to the lightning rod conductors. This rule will prevent flashovers in the event of a lightning stroke.

An FPN refers to the latest issue of NFPA 780, *Lightning Protection Code* for more information on the subject.

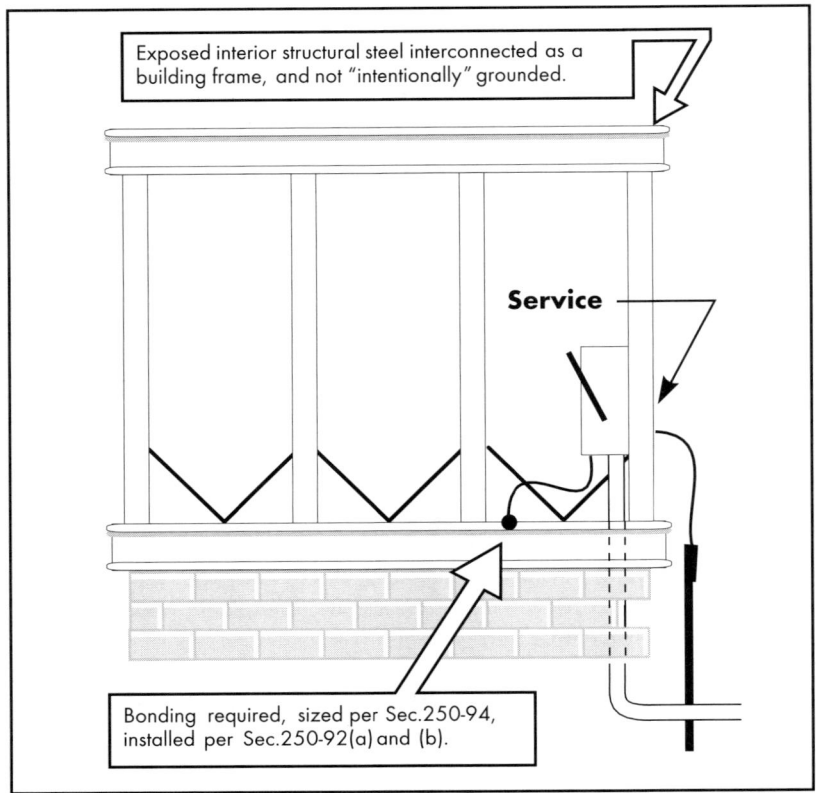

Exposed interior structural steel interconnected as a building frame, and not "intentionally" grounded.

Service

Bonding required, sized per Sec.250-94, installed per Sec.250-92(a) and (b).

Fig. 3.8 Bonding also applies to exposed interior structural steel.

4

ARTICLE 250, PART B — CIRCUIT AND SYSTEM GROUNDING

Circuit and system grounding, as explained in Chapter 1 of this book, basically involves connecting part of an AC power distribution system to ground. The subject, however, is complex. Even in a simple configuration in which the star point of transformer wye-connected secondaries is grounded, the system can be either solidly grounded or impedance grounded. But wye-connected secondaries are not the only ones that can be grounded; since there are also different ways in which a delta-connected secondary can be grounded.

The rules also vary on *where* systems are to be grounded. Generally, systems are grounded at the service entrance equipment. Within an outbuilding of a multiple building facility, the incoming distribution system can be grounded locally. Also, DC systems must comply with regulations on grounding.

NEC provisions for circuit and system grounding, therefore, are divided into several parts. Part B lists the types of systems that must be grounded, with the exception of dc systems. The NEC now groups all dc systems in a new Part H, however, we will continue to integrate this material as we go along.

SEC. 250-20 – AC CIRCUITS AND SYSTEMS TO BE GROUNDED

As detailed in the following subsections, some AC systems are required to be grounded. Other AC systems *may* be grounded even if not required to be grounded. For example, a FPN mentions a delta-connected transformer that has been corner grounded as an example of a system where grounding is permitted even though the Code does not require it to be grounded.

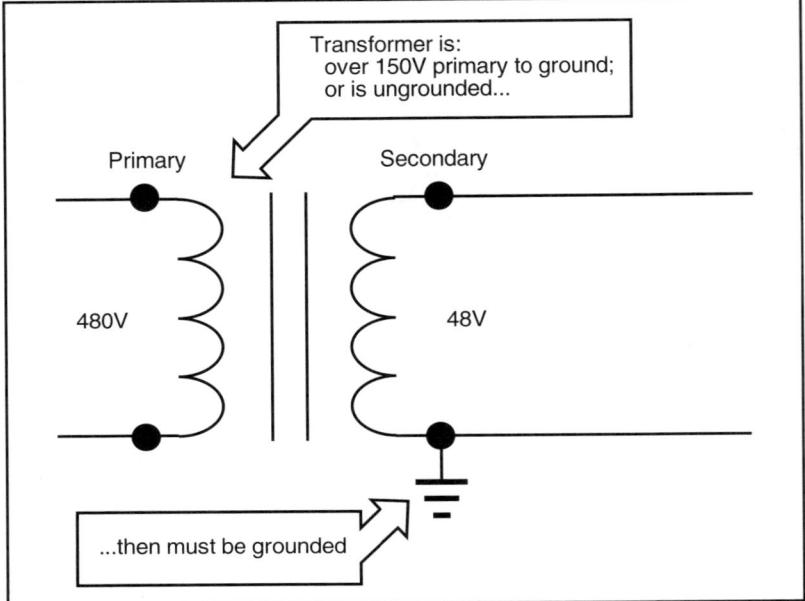

Fig. 4.1 The general rule for grounding AC systems supplying circuits operating at less than 50V.

(a) **AC circuits of less than 50V** must be grounded (as shown in **Fig 4.1**) if:

• They are supplied by a transformer whose primary is supplied at over 150V to ground; or

• The system supplying the transformer primary is ungrounded; or

• They are installed as overhead conductors outside of buildings.

(b) **AC circuits of 50V to 1000V.** There are similar rules for AC systems operating between 50 and 1000V that supply premises wiring. So it is first necessary to understand what the Code means when it refers to premises wiring.

As defined in Article 100 in the NE Code, premises wiring is interior and exterior power, lighting, control, signal, and other circuit wiring that extends from the service point of utility conductors, or other source of power such as a battery, or a solar photovoltaic system, or from genera-

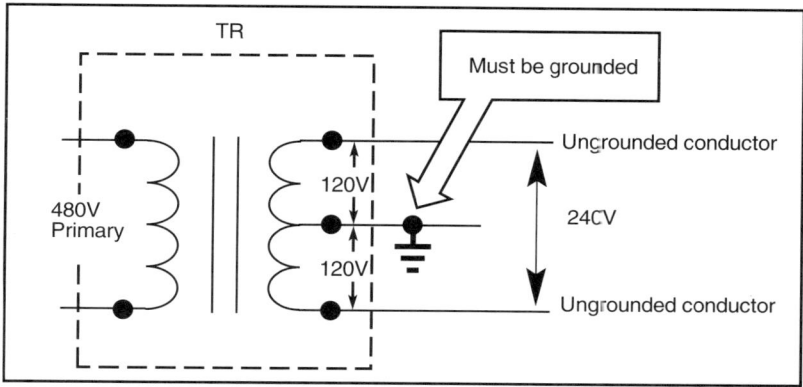

Fig. 4.2 An AC system operating from 50V to 1000V must be grounded if it is arranged so that the voltage to ground is less than 150V.

tor, converter, or transformer windings to the outlets. Premises wiring also includes all of the associated hardware, fittings, and wiring devices enclosing the wiring system. Temporary wiring as well as permanently installed wiring are covered by this definition.

Conductors of devices plugged into receptacles are not considered to be premises wiring. The internal wiring of appliances, lighting fixtures, motors, motor controllers, motor control centers, and similar equipment, are also not considered to be premises wiring.

An AC system that supplies premises wiring and has an operating voltage between 50V and 1000V *must* be grounded:

• Where the system can be grounded so that the maximum voltage to ground on the ungrounded conductors does not exceed 150 volts to ground (see **Fig. 4.2**); or

• The system is 3-phase, 4-wire, wye-connected in which the neutral is used as a circuit conductor (see **Fig. 4.3A**), or

• The system is 3-phase, 4-wire, delta connected in which the mid-point of one phase is used as a circuit conductor (see **Fig. 4.3B**).

Sec. 250-21 lists many exceptions to the requirement for grounding AC systems operating at between 50V and 1000V:

(1) The general rule does not apply if the system is used exclusively to supply industrial furnaces for melting, refining, tempering, or similar operations.

(2) The general rule does not apply for separately-derived systems used exclusively for rectifiers supplying adjustable-speed drives

(3) The general rule does not apply where the transformer is fed at less than 1000V, continuity of power is essential, and the system exclu-

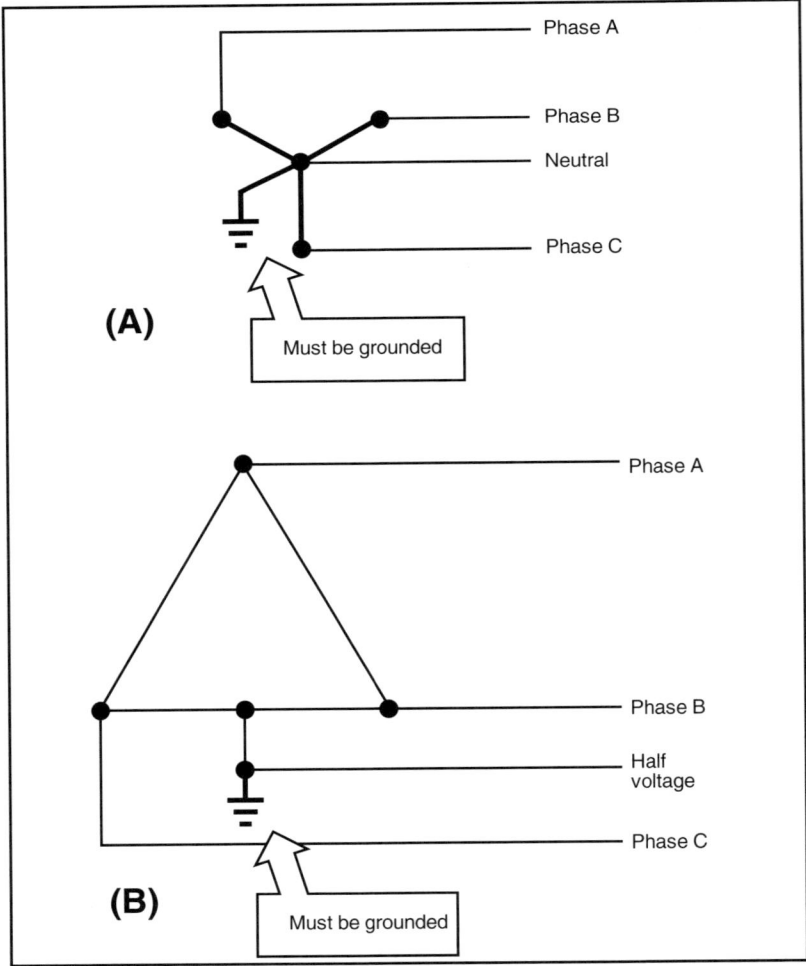

Fig. 4.3 Both Wye–connected systems **(A)** and delta-connected systems where a phase winding is tapped **(B)** to connect a circuit conductor must be grounded.

sively supplies critical control circuits. To apply this exception, the system must be maintained by qualified persons only, and ground-detectors must be installed on the control system.

(4) The general rule does not apply for isolated systems in health care facilities permitted in Article 517, and electrolytic cells as permitted in Article 668. In these cases, an FPN suggests that ground detectors will enhance safety.

(5) The general rule does not apply where (see **Fig 4.4**) the system employs a high-impedance grounded neutral in which a grounding impedance (usually a resistor) is used to limit the ground-fault current to a low value, as detailed in Sec. 250-36. These systems are permitted for 3-phase AC systems operating at between 480V and 1000V, provided that

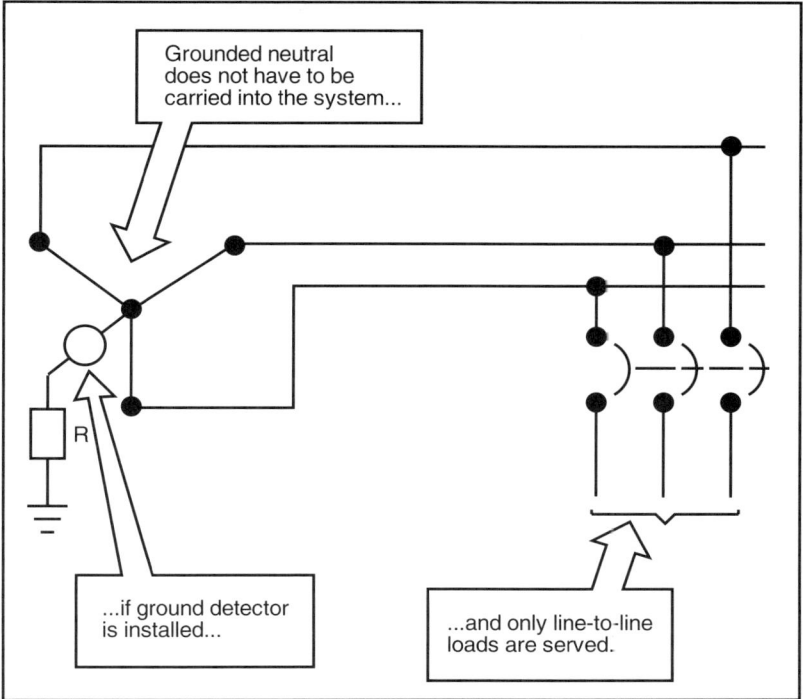

Fig. 4.4 An exception to system grounding requirements is made for systems employing high-impedance grounding.

continuity of service is essential, and line-to-neutral loads are not served. Only qualified persons are permitted to maintain these installations, and ground-detectors must be installed within the system. These systems are covered in this book in more detail in Chapter 5.

(c) **AC systems of 1kV and greater** that supply mobile or portable equipment must also be grounded. Other AC systems operating at 1000V or more are permitted to be grounded, but are not required to be.

(d) **Separately derived systems.** Art. 100 defines a separately derived system as a premises wiring system whose power is derived from a battery, or a solar photovoltaic system, or from a generator, transformer, or converter windings, and that has no direct electrical connection, including a solidly connected grounded circuit conductor, to supply conductors

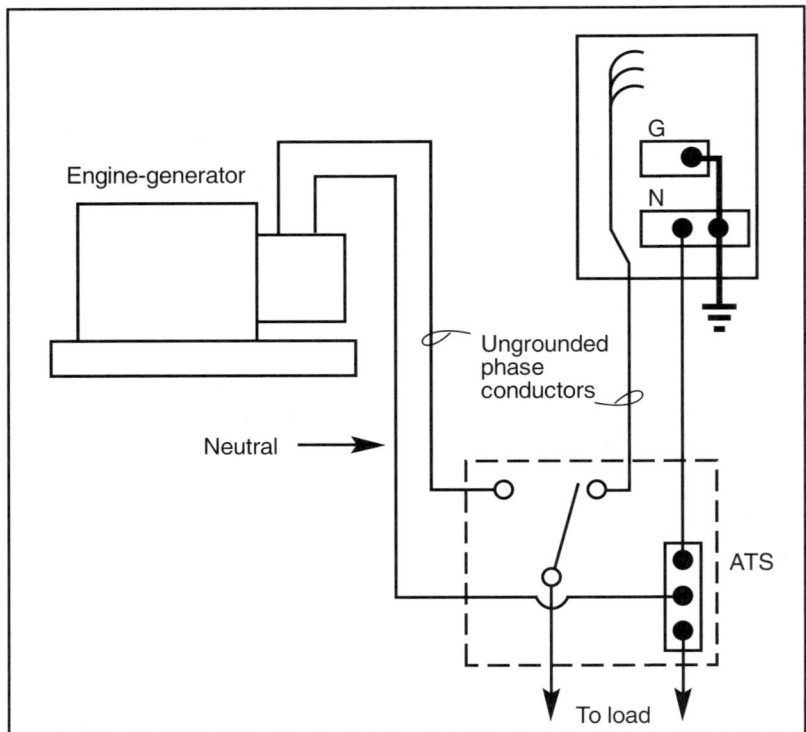

Fig. 4.5 This is not considered to be a separately-derived system.

originating in another system.

The system grounding rules apply unless the system supplied by the separately derived system is not one that requires grounding under the rules of Sec 250-20. For example, a delta-connected 480V, 3-phase, separately derived system need not (but could) be grounded.

Two FPN's address an AC power source such as an on-site generator whose neutral is solidly interconnected to a service-supplied neutral (as shown in **Fig. 4.5**). It *is not* considered to be a separately-derived system. In this event, conductors that must carry ground-fault currents are required to be not smaller than specified in Sec 445-5 The sizing is discussed further in Chapter 11 of this book.

SEC. 250-22–CIRCUITS NOT TO BE GROUNDED

The Code makes a specific statement that circuits must *not* be grounded in the following installations:

• Cranes operating over combustible fibers in Class III locations, per Sec. 503-13.

• Health-care facilities as required in Article 517 [Sec. 517-160(a)(2)].

• Electrolytic cells as provided in Article 668 [Sec. 668-11(a); 668-21(a)].

SEC. 250-26–AC SYSTEM CONDUCTOR TO BE GROUNDED

For an AC premises wiring system, the following conductor is to be grounded and identified as required in Article 200:

• One conductor of a single-phase, 2-wire system.

• The neutral of a single-phase, 3-wire system.

• The common conductor of a multiphase system having one wire common to all phases, such as a 3-phase, 4-wire wye-connected system.

• One phase conductor of a multiphase system requiring one grounded phase, such as a corner-grounded delta system.

• The conductor connected to the center tap of a multiphase system in which one phase is used as a single-phase, 3-wire system (such as a 3-phase, 3-wire delta system where one phase is center tapped).

SEC. 250-34–PORTABLE AND VEHICLE-MOUNTED GENERATORS

A special set of rules for grounding apply to portable equipment. They are given in the following subsections.

(a)Portable generators. The frame of a portable generator is not required to be grounded and is permitted to serve as the grounding electrode for a system supplied by the generator if:

• the generator supplies only equipment mounted on the generator or cord-and-plug connected equipment through receptacles mounted on the generator, or both; or

• the noncurrent-carrying metal parts of equipment and the equipment grounding conductor terminals of the receptacles are bonded to the generator frame.

(b) Vehicle-mounted generators. The frame of a vehicle is permitted

Portable generators must meet NEC grounding rules.

to serve as the grounding electrode for a system supplied by a generator mounted on the vehicle provided that:

- the frame of the generator is bonded to the vehicle frame; and
- the generator supplies only equipment located on the vehicle or cord-and-plug-connected equipment through receptacles mounted on the vehicle, or both equipment located on the vehicle and cord-and-plug-connected through receptacles mounted on the vehicle or on the generator; and
- the noncurrent-carrying metal parts of equipment and the equipment grounding conductor terminals of the receptacles are bonded to the generator frame; and
- the system complies with all other provisions of Article 250.

(c) Grounded conductor bonding. A system conductor that must be grounded in accordance with Sec. 250-26 must be bonded to the generator frame where the generator is a component of a separately derived system. In addition, under some conditions an equipment grounding conductor will be bonded to the generator frame. Note that the state-

Truck-mounted generators covered by NEC rules on grounding and bonding include those used for welding.

ment in prior codes that no conductor other than a neutral need be bonded to the generator frame has been deleted from the 1999 NEC.

An FPN refers to Sec. 250-20(d) for grounding of portable generators supplying fixed wiring systems. This is because such generators may very possibly become separately derived systems depending on transfer switch design and other considerations.

SEC. 250-162–DC SYSTEMS

Grounding of DC systems is required at the higher voltages because they present as much danger to personnel as AC systems. The DC systems that are required to be grounded are described in the following subsections.

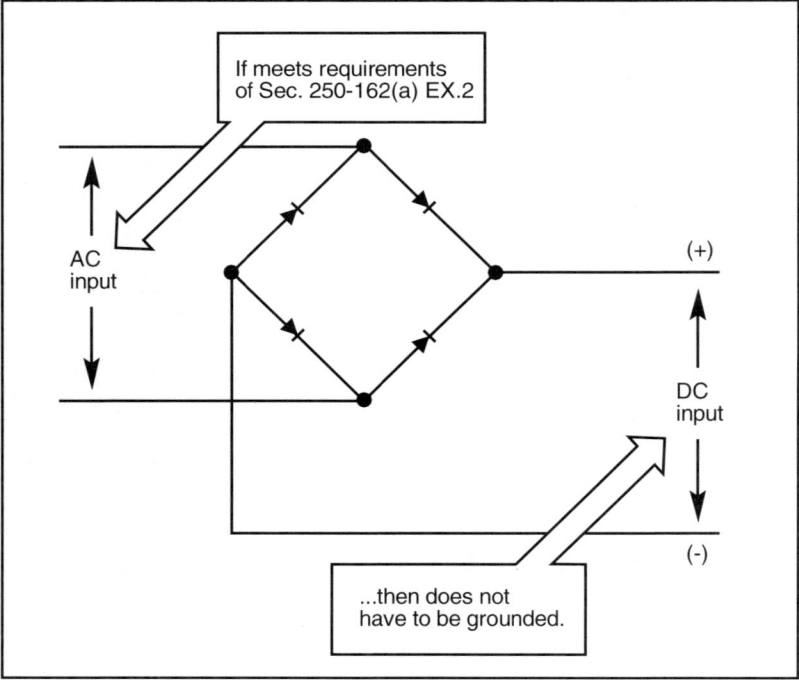

Fig. 4.6 One of the exceptions to the requirement that 2-wire DC systems must be grounded.

(a) Two-wire DC systems. The general rule is that two-wire DC systems supplying premises wiring must be grounded. But there are many exceptions given to this rule. The 2-wire system is not required to be grounded if it:

• Supplies only industrial equipment in a limited area and a ground detector is installed; or

• Operates at 50 V or less between conductors; or

• Operates at over 300V between conductors; or

• Is derived from a rectifier that is supplied from an AC system that meets the Code requirements for system grounding (see **Fig. 4.6**); or

• Is part of a power-limited DC fire protective signaling circuit having a maximum current of 0.030 A.

(b) Three wire DC systems that supply premises wiring must have the neutral conductor grounded (see **Fig. 4.7**). There are no exceptions granted for this rule.

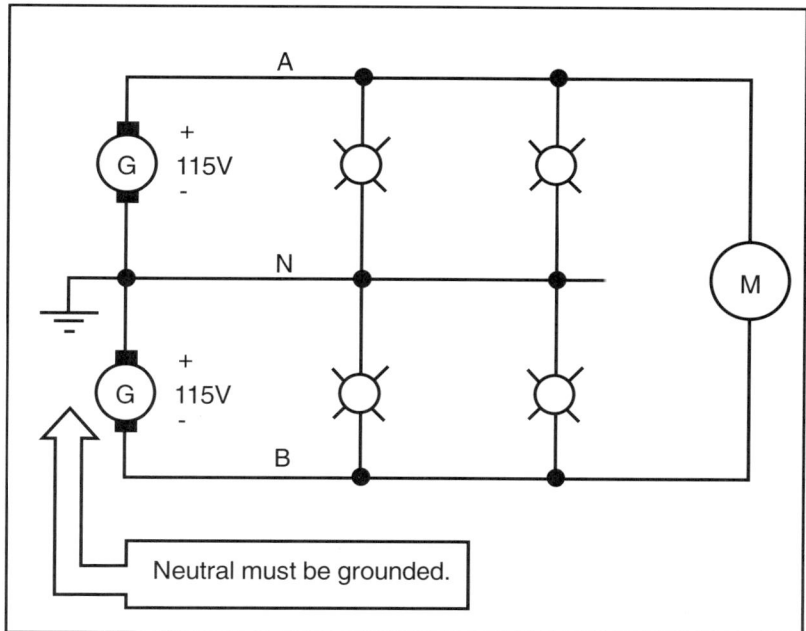

Fig.4.7 The general rule for grounding a typical 3-wire DC system.

ARTICLE 250, PART B CON'T— SYSTEM GROUNDING CONNECTIONS - A TECHNICAL REVIEW

5

This chapter further extends the rules on system and circuit grounding to cover how and where the connections are to be made. These rules are taken from selected parts of Article 250, grouped together here to explain in one chapter the required locations for system grounding connections and the technical reasons these requirements exist.

Never forget the fundamental reason for grounding electrical systems, as given in Sec. 250-2(a):

(a) Grounding of Electrical Systems. Electrical systems that are required to be grounded shall be connected to earth in a manner that will limit voltages imposed by lightning, line surges, or unintentional contact with higher voltage lines, and that will stabilize the voltage to earth during normal operation.

In addition, a grounded system provides for automatic clearing of ground faults. In grounded systems, equipment grounding conductors are bonded to the system grounded conductor to provide a low impedance path for fault current that will facilitate the operation of overcurrent devices under ground-fault conditions. There are, however, cases where this must be modified.

SEC. 250-24–GROUNDING SERVICE-SUPPLIED AC SYSTEMS

A large majority of facilities receive their electric power from a serving utility. This can be at a voltage level that is directly usable by the customer, as in a typical single-family residence, or at a higher voltage that is later stepped down by the customer within the site, as is done in a large manufacturing facility. The rules for grounding contained in this section apply totally in the first instance, and up to the point of the voltage transformation in the second, beyond which the rules of Sec. 250-30 apply if the secondary qualifies as "separately derived."

The following subsections detail the NEC requirements for grounding of a service-derived AC system.

(a) System grounding connections. A premises wiring system supplied by an AC service that is required to be grounded (as defined in Sec. 250-20 or 250-21) must have at each service (**Fig. 5.1**) a grounding electrode conductor connected to a grounding electrode of the permitted type. These are described in detail in Chapter 2 of this book. They include:

- a metal underground water pipe;
- the effectively grounded metal frame of a building;
- a concrete-encased electrode; or
- a ground ring.

When none of the above preferred electrodes are available, a made electrode is acceptable. This can be a ground rod or other type of made electrode described in Sec. 250-52 (previously covered in Chapter 2 of this book).

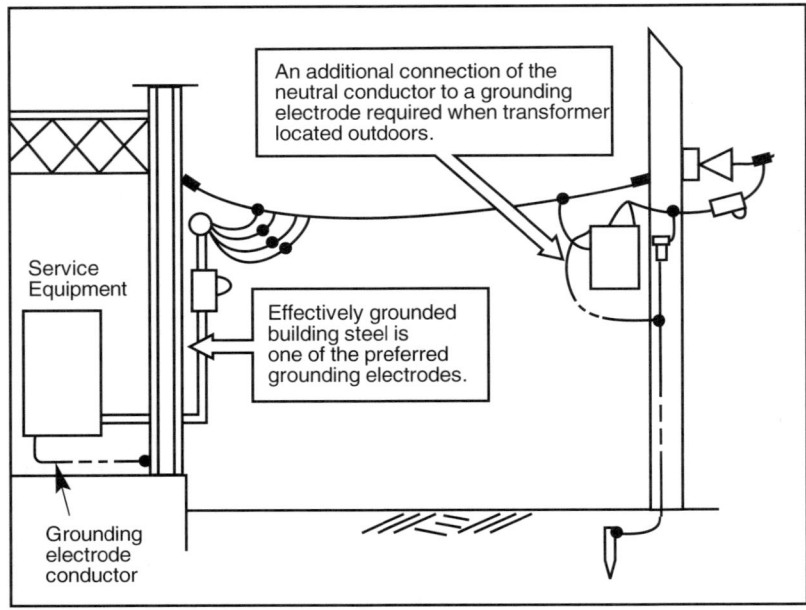

Fig. 5.1 Typical grounding of a service entrance.

The general rule is that the grounding electrode conductor must be connected to the grounded service conductor at any accessible point from the load end of the service drop or lateral, to and including the bus or terminal to which the grounded service conductor is connected at the service disconnecting means. There are modifications and refinements to this general rule, however, as follows:

(1) When a transformer is used to supply the service, and the transformer is located outside the building, at least one additional grounding connection must be made from the grounded circuit conductor to a grounding electrode, either at the transformer, or elsewhere outside the building. You must never make this connection, however, if you are supplying a high-impedance grounded neutral system. If you do, you will desensitize the fault sensor by allowing ground return currents to bypass it and flow through the earth to the remote electrode.

(2) Where a double-ended (dual-fed service) switchboard (**Fig. 5.2**) is in a common enclosure, and has a secondary tie, a single grounding

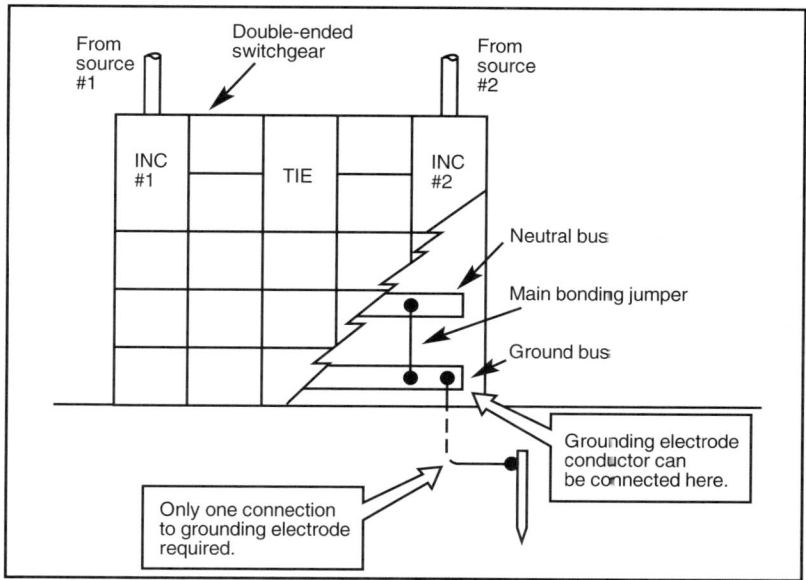

Fig. 5.2 Exceptions apply to the rule on grounding electrode conductors for dual-fed services, and where the main bonding jumper is a wire or busbar.

electrode connection to the tie point of the grounded circuit conductors is permitted instead of two separate grounding electrode connections.

(3) Where the service equipment contains both a neutral and a ground bar or bus, and a bonding jumper consisting of a busbar or a wire (but not a screw) (**Fig. 5.2**) that connects the two together, the grounding electrode conductor can be connected to the ground bar or bus, or to the neutral bar or bus. Some arrangements to sense ground faults require the GEC to be connected to the grounding bus, because they measure fault return currents over the main bonding jumper. Such a sensor must sense all equipment grounding conductor return fault current, with none bypassing the sensor through random passages through earth or other grounded objects to the grounding electrode.

(4) Making a grounding connection to the grounded circuit conductor (neutral) on the load side of the service disconnect as shown in **Fig. 5.3** is prohibited under most circumstances. The FPN notes that exceptions to this rule are:

• As shown in Sec. 250-30(b), a grounding electrode conductor must be connected to the grounded (neutral) conductor of downstream equip-

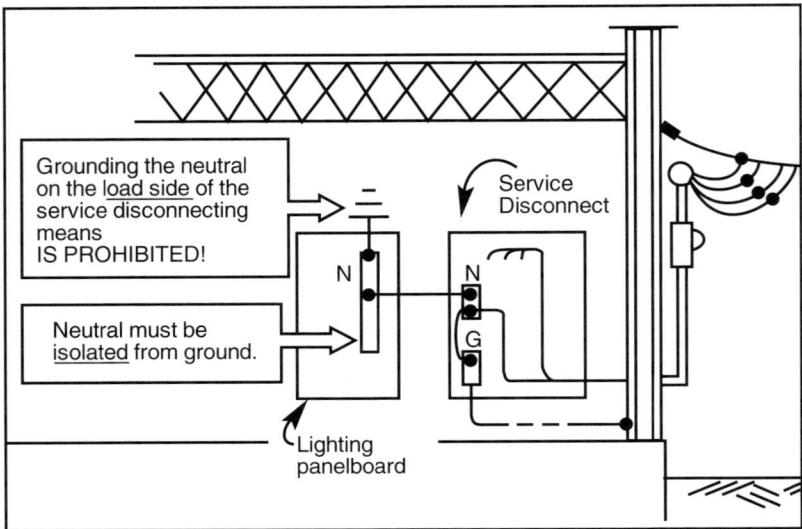

Fig. 5.3 Grounding of the neutral in this case is not allowed.

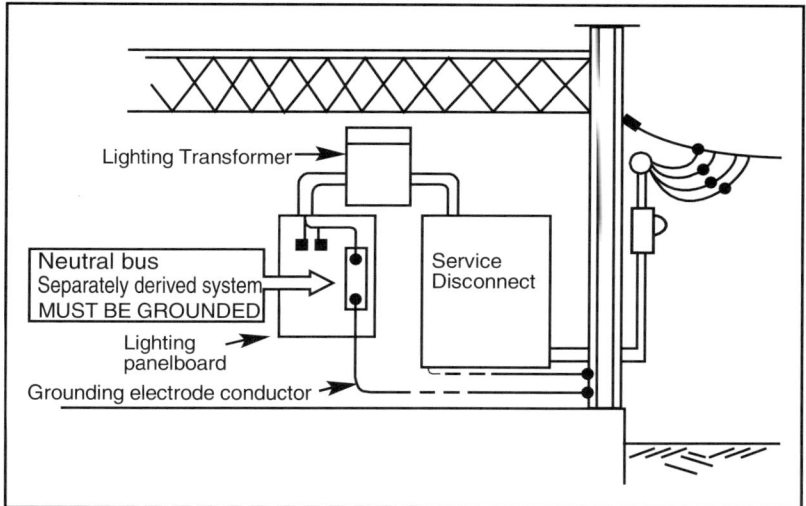

Fig. 5.4 Grounding of the neutral in this case is mandatory.

ment that is energized from a separately-derived system as shown in **Fig. 5.4**.

   • As shown in 250-32 for connections at separate buildings or structures
   • As shown in 250-142 for use of the grounded circuit conductor to ground equipment.

(b) Grounded conductor brought to service equipment. When an AC system operating at 1000V or less is grounded at any point, the grounded conductor(s) must be run to the service disconnect along with the phase conductors and bonded to the disconnect enclosure (see **Fig. 5.5**). This requirement is valid whether or not the neutral is to be used in the distribution system of the facility, since the neutral conductor is the ground-fault current return path to the transformer neutral point.

   Sec. 250-24(b)(1) requires that the grounded conductor brought to the service equipment must be run to each service disconnecting means and be bonded to each disconnecting means enclosure, and that it be routed with the supply phase conductors.

   Sec. 250-24(b)(2) refers to Sec. 310-4 for rules that apply to grounded conductors connected in parallel.

   The sizing of grounded conductors is discussed in Chapter 11 of this book.

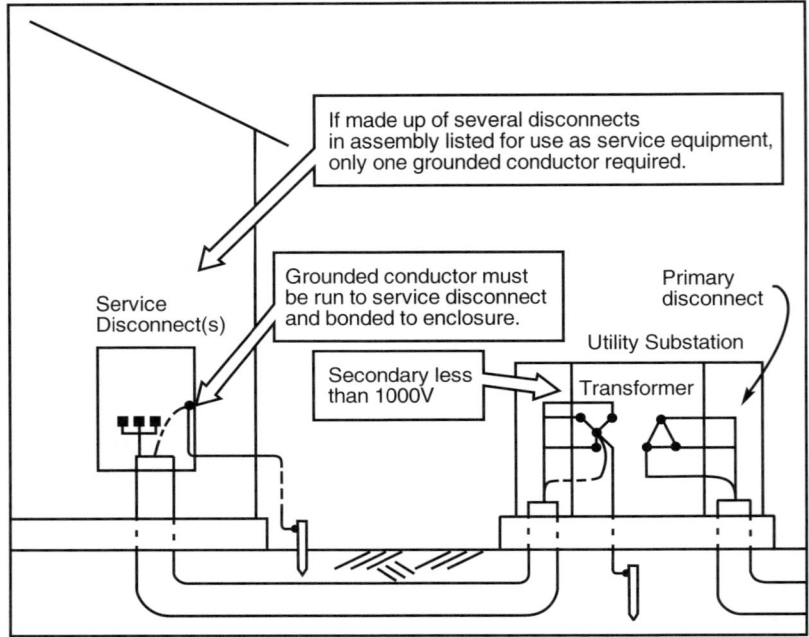

Fig. 5.5 The grounded (neutral) conductor of a system operating at 1000V or less must be brought to the service equipment.

Sec. 250-24(b), Exception states that if multiple disconnects serve as the service disconnects, and they are located in an assembly listed for use as service equipment, then only one grounded conductor is required to be run to the assembly and be bonded to the enclosure.

Sec. 250-24(b)(3) coordinates this rule with the requirements of Sec. 250-36, which is treated in the discussion of that section. There are good reasons to apply these rules once the system is grounded at *any* point, even if you aren't using line-to-neutral loads.

Referring to **Fig. 5.6**, suppose a power load supplied from Service Enclosure No. 1 develops a ground fault. Whether or not a grounded conductor enters the enclosure, the system is still grounded, and it will attempt to behave accordingly. The fault return current will still seek out the transformer neutral point to complete the circuit, probably through Service Enclosure No. 2. Current would flow over the two grounding electrode

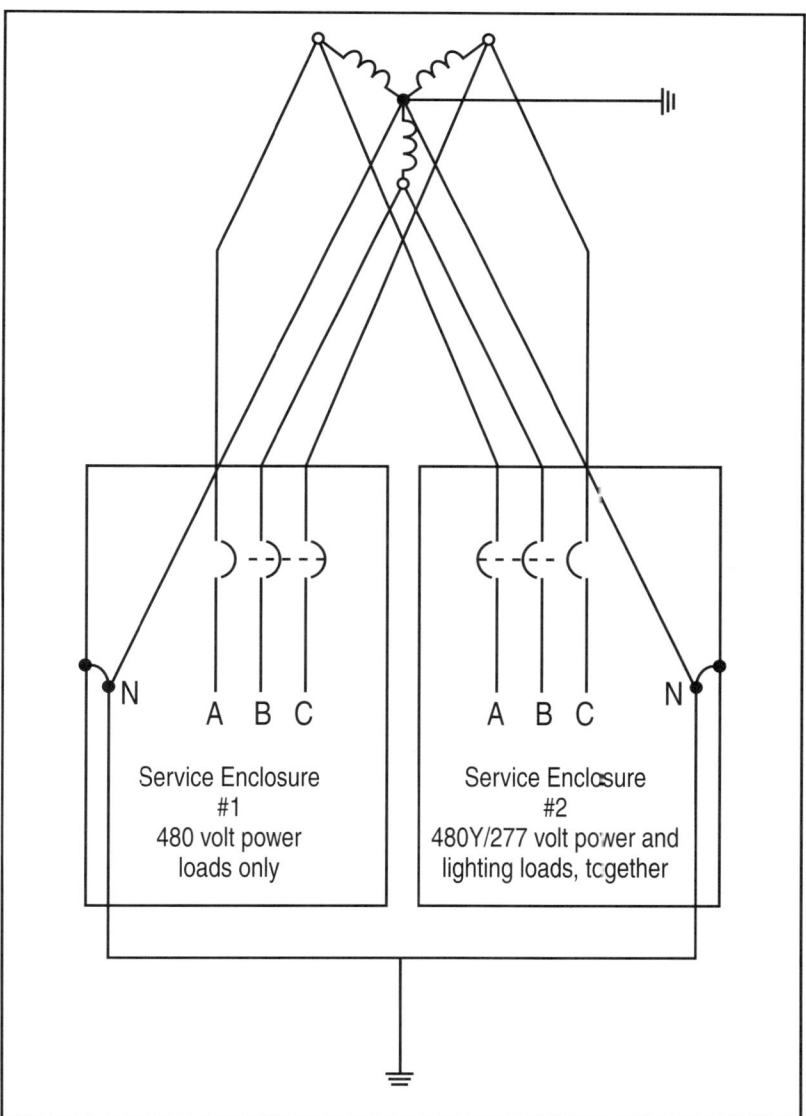

Fig. 5.6 A grounded circuit conductor, large enough for the load and at least large enough to adequately return fault currents, must enter both enclosures.

conductors and through part of the grounding electrode, although with uncertain results given that the path was never designed for this duty.

The fault wouldn't clear quickly as a result, and also because of the increased reactance that comes from the return path being at a significant distance from the supply conductors. The result is the worst of both worlds - an "ungrounded" system with no ability to peacefully ride through an initial fault without an outage, and a "grounded" system without the ability to properly clear the same initial fault.

SEC. 250-28 – MAIN BONDING JUMPER

In a grounded system, this conductor is arguably the most important single conductor of the entire system. This is the conductor that makes the grounded system work, for without it, any fault current in an equipment grounding conductor that returns (and thereby makes a complete circuit) to its source is essentially accidental. Within the service disconnecting means enclosure, this conductor connects together all the equipment grounding conductors supplied by that service, and connects them to the grounded circuit conductor. Remember, electrical systems supply conductors, not raceways (**Fig. 5.7**).

The size follows similar requirements to those for Sec. 250-24(b)(1), and we cover this in Chapter 11. It must be attached per Sec. 250-8, and, if a screw, must be colored green. This allows inspectors to quickly determine if it has been installed in service panelboards, for example. These have numerous screws through the grounded conductor bus, many of which dead-end into nonmetallic insulating supports. The green marking is distinctive. Nevertheless, inspectors often report these bonding screws left loose in the bottom of panels, thereby omitting the most important single conductor of the entire wiring system.

SEC. 250-30 – GROUNDING SEPARATELY DERIVED AC SYSTEMS

Sec. 250-30 describes how separately derived systems must be grounded. It is divided into subsections (a) and (b), covering grounded and ungrounded systems. We'll cover the numbered paragraphs in (a) on grounded systems first.

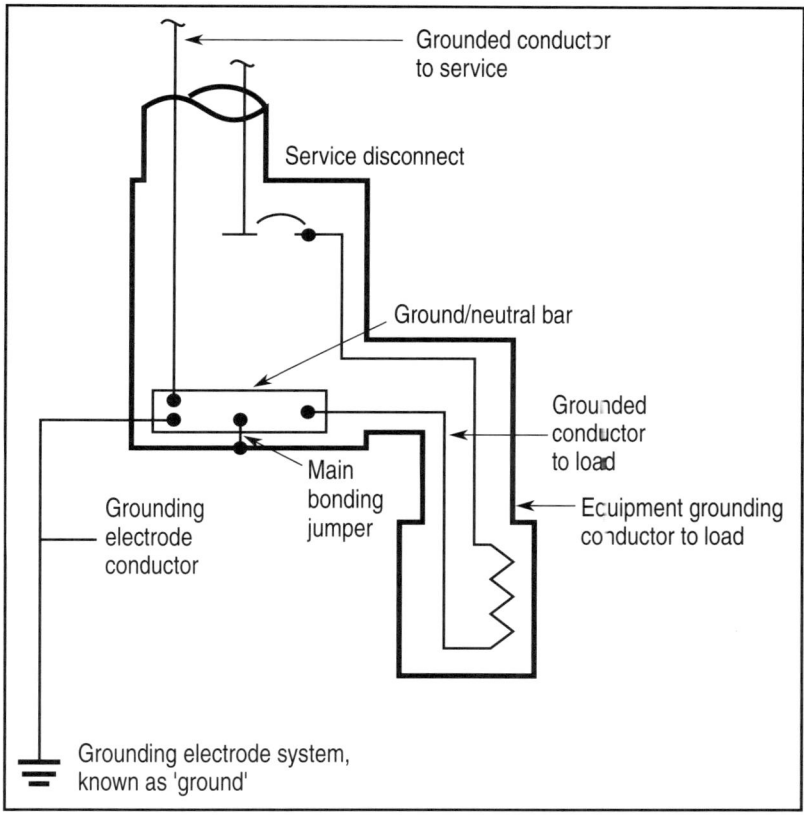

Fig. 5.7 The main bonding jumper connects all equipment grounding conductors to a grounded circuit conductor at the service.

(1) Bonding jumper. A properly-sized bonding jumper must be used to connect the equipment grounding conductors of the derived system to the grounded conductor. This connection must be made at some point between the source and the first system disconnect means or overcurrent device of the derived system. If there is no overcurrent device beyond the source, bonding between the grounding and grounded conductor must be made at the source.

Be aware of a significant error in the 1999 NEC at this point. Sec. 250-30(a)(1) now refers to Sec. 250-24(a)(4), a provision that has nothing to do with the location of the bonding jumper. Before this Code edition, and all the way back to the 1975 edition, this provision referred to what is now Sec. 250-24(a)(3) on double-ended services. This accommodated selective ground-fault protection of equipment (GFPE) schemes that are available to permit dual-source systems to operate with maximum service continuity. Under these systems, a ground fault only disables half the loads, with the other side remaining operable.

This change in the 1999 NEC is a completely unsubstantiated major change that effectively disallows a downstream location for the bonding jumper (and the omission of all other secondary neutral-to-ground connections) in cases where separately derived systems are arranged in a double-ended design with a center tie point. Although Sec. 250-30(a)(2) allows for the grounding electrode conductor to be connected at such a point, *the bonding jumper must be connected at the same point in order to avoid parallel current through enclosures.* Further, if the bonding jumper is connected at another location, the selective GFPE arrangements won't work correctly. This will need to be corrected in the next Code cycle.

Sizing of the bonding jumper is detailed in Chapter 11 of this book. There are two exceptions to Sec. 250-30(a)(1):

(1) Two bonding jumpers (one at the source and one at the first disconnecting means) may be installed where doing so does not establish a parallel path for the grounded circuit conductor. This essentially parallels the rule for second buildings in Sec. 250-32(b)(2), and for the same reason. The Code, to the extent practicable, is trying to assure that grounded circuit conductor power currents stay confined to recognized electrical conductors. Note that a connection through the earth is not considered a parallel path, and won't defeat this exception.

The earth is the only exemption here. For example, suppose a short run of greenfield connects a separately derived system transformer and a panel. You need a separate bonding conductor for the greenfield; you can't simply bond both the panel and the transformer ends to the grounded circuit conductor because so doing would create this parallel path.

(2) The size of bonding jumper for an Article 725 circuit (Class 1, 2, or 3) derived from a small (not larger than 1000VA) transformer may be smaller than is shown in Table 250-66, but not smaller than the phase conductors energizing the transformer and not smaller than No. 14 cop-

per or No. 12 aluminum. This effectively allows a supply equipment grounding conductor to function as a grounding electrode conductor. It is very common in combination controllers with control transformers. Note carefully, that *this is also the only allowance* for the equipment grounding conductor of the supply wiring method to function in this way.

(2) Grounding electrode conductor. In addition to connecting together the equipment grounding conductor and the grounded circuit conductor at a separately-derived source, a grounding electrode conductor must also be connected at some point in the separately-derived system. This connection can be anywhere from the source to the first dis-

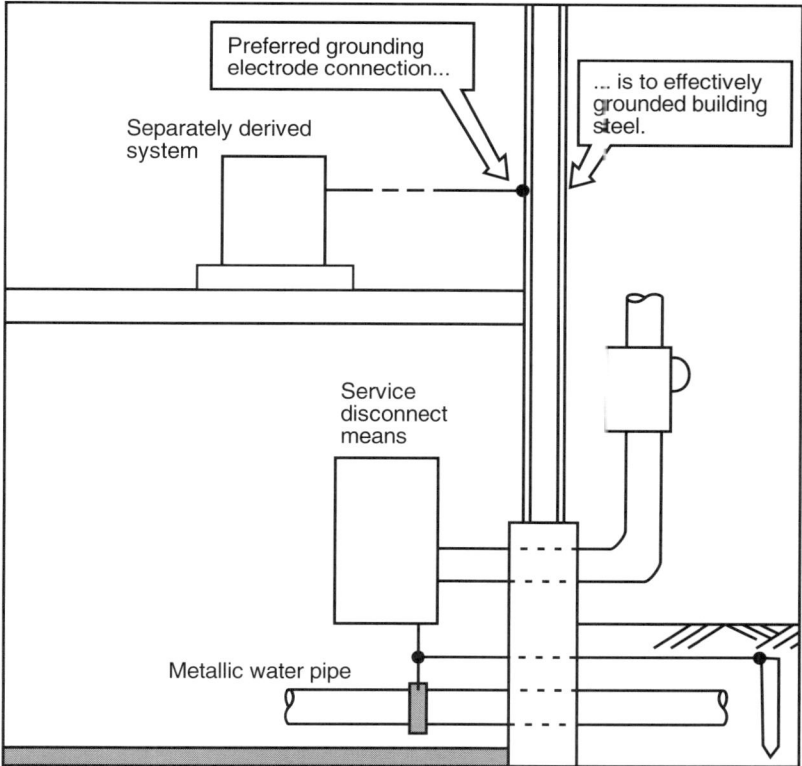

Fig. 5.8 A separately derived system must be connected to the closest preferred grounding electrode.

connecting means or overcurrent device; however it must be made to the same point where the bonding jumper is installed except for installations of double-ended dual-fed supplies [Sec. 250-24(a)(3)] or where the main bonding jumper is a wire or busbar in the supply equipment [Sec 250-24(a)(4)]. Similar to the requirements for the bonding jumper given in (1), if the separately derived system has no disconnecting means or overcurrent devices, the connection must be made at the source.

Sizing of the grounding electrode conductor is covered in Chapter 11 of this book.

(3) Grounding electrode. The grounding electrode to which the grounding electrode conductor from the separately-derived system is connected must be either the nearest available effectively grounded structural metal member of the structure, or the nearest available effectively grounded metal water pipe. You must make this connection as near as practicable to, and preferably in the same area as, the grounding connec-

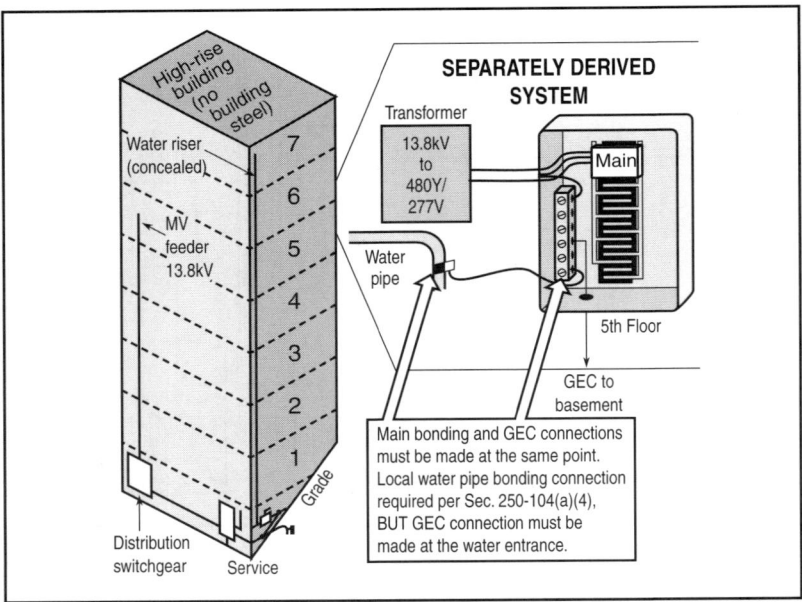

Fig. 5.9 There will be a very long grounding electrode conductor required, from the separately derived system on the 5th floor to the basement.

tion to the separately derived system.

This Code section makes clear that the grounding electrode must be as near as practicable to the separately-derived source, preferably in the same area as the grounding electrode conductor connection to the system. As seen in **Fig. 5.8**, the building column would be preferred to the water pipe in this case because it is closer to the separately derived system.

If you have your eye on the local water pipe as being closer, remember it must also meet the 5-ft-into-the-building restriction as water pipe electrodes generally have to meet (and with the same available exception) as in Sec. 250-50 (**Fig. 5.9**). This was discussed in Chapter 2 of this book.

Only if either of these is not available can other recognized grounding electrodes be used. There is an exception to these rules that addresses the common situation where a separately derived system originates in a listed substation, as illustrated in **Fig. 5.10**.

(4) Grounding methods. All of the methods used for achieving ground-

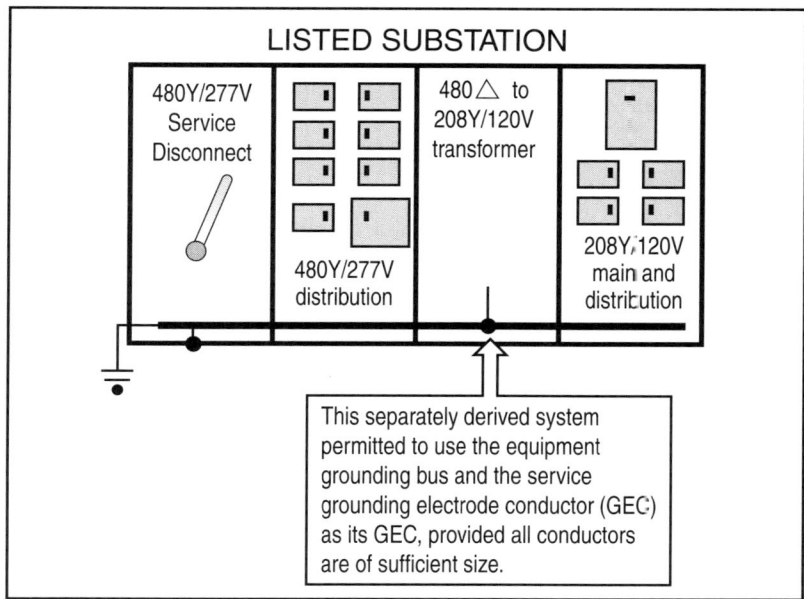

Fig. 5.10 This unit substation is a common source for separately derived systems. If listed, its ground bus can perform double duty.

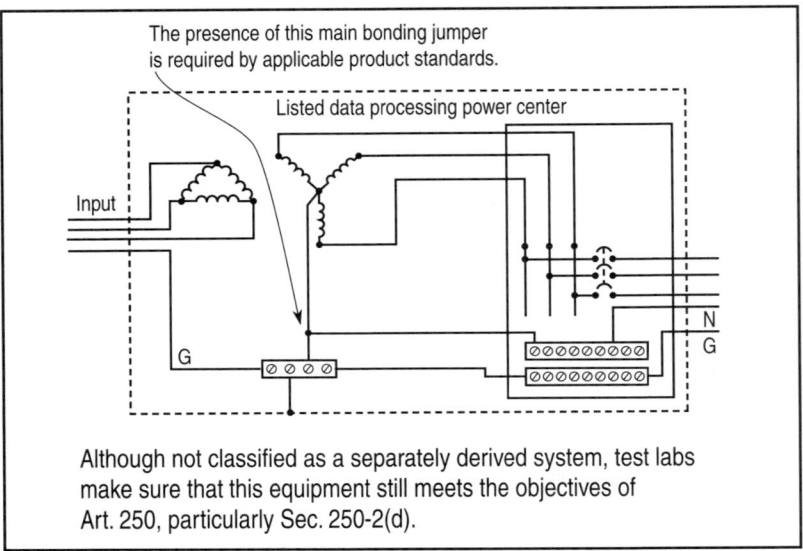

The presence of this main bonding jumper is required by applicable product standards.

Listed data processing power center

Input

G

N

G

Although not classified as a separately derived system, test labs make sure that this equipment still meets the objectives of Art. 250, particularly Sec. 250-2(d).

Fig. 5.11 Sec. 645-15 and related (FPN No. 1) address grounding in information technology rooms. Note that, per sec. 645-1, the provisions of Art. 645 don't extend beyond the boundaries of such a room.

ing mandated by this section must comply with the requirements of the other parts of the NEC.

Information technology equipment. One of the more controversial provisions of the NEC over the years has been "separately derived systems" that Sec. 645-15 says, for Code enforcement purposes, aren't to be considered that way (see **Fig. 5.11**). These systems can cover extensive areas, and now Sec. 645-15 requires signal reference structures, if installed, to be bonded to the equipment grounding system supplying the equipment. These power supply systems must be listed, and Sec. 645-15 (FPN No. 1) essentially assures that the testing labs will verify compliance with the objectives of Art. 250.

Ungrounded systems: Sec. 250-30(b) provides parallel rules for ungrounded separately derived systems to those that apply to grounded systems. There are a few differences, the most important being that the grounding electrode conductor goes to either the system disconnect enclosure, or to the source enclosure, or to any enclosure between those two points.

Since the system is ungrounded, subjecting the external enclosures or grounded objects to the routine flow of circuit current isn't a problem. Except as allowed for portable and vehicle-mounted generators in Sec. 250-34, the electrode must meet the requirements in Sec. 250-30(a)(3). All other requirements are the same as for grounded systems.

Sec. 250-169 imposes similar requirements on ungrounded dc systems. There is the same allowance for the portable or vehicle-mounted generator, and for the point of connection between the grounding electrode conductor and the wiring system. The size follows similar rules for nonseparately derived dc systems in Sec. 250-166, which we cover in Chapter 11.

SEC. 250-32–TWO OR MORE BUILDINGS OR STRUCTURES SUPPLIED FROM A COMMON SERVICE

It is quite common for a large industrial, commercial, or institutional facility to be made up of many buildings within a campus or plant site. Often, these complexes receive power from one or more of the serving electrical utility's substations(s) located within, or immediately adjacent to the site. From that point, either the utility or the customer distributes

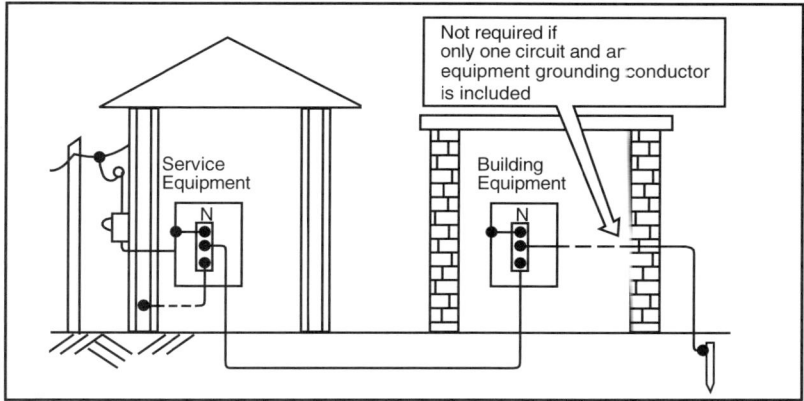

Fig. 5.12 Grounding of building subfed from grounded system located in another building.

the power to the separate buildings at the voltage received, or can transform it to a lower voltage and then distribute it, or can transform the voltage at unit substations at convenient locations within the facility, or can subfeed one or more buildings from a single service entrance. In every case, however, there will be a local grounding electrode established at each building or structure. The only exception to this is the trivial case where only one branch circuit is involved (**Fig. 5.12**). The rules that apply in most cases are detailed in the following subsections and numbered paragraphs.

(1) **Equipment grounding conductor.** This rule applies to situations where buildings are subfed from service equipment located in a separate building. As shown in **Fig. 5.13**, a grounding electrode must be established not only at the main service disconnect at the first building of supply, but also at each building or structure subfed from the service, and then connected to the supply equipment grounding conductor.

The system grounded (neutral) conductor must not be connected to the equipment grounding conductor or to the grounding electrode(s) at the separate subfed building. From a practical perspective, this treats the

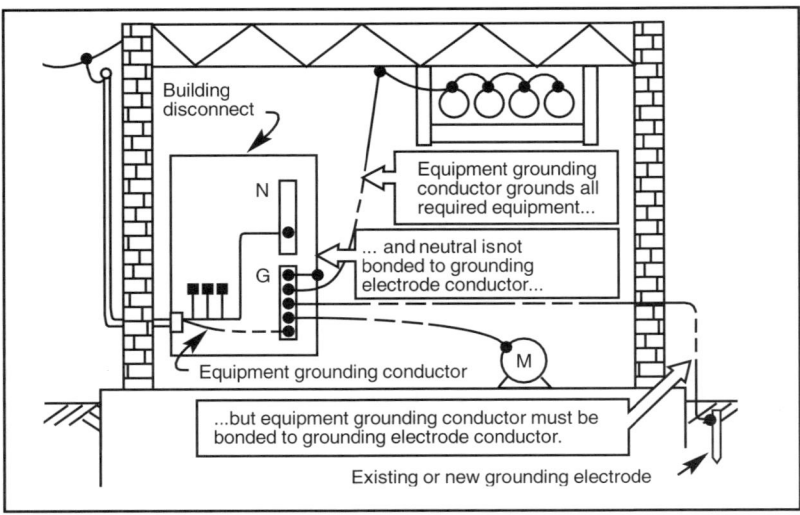

Building disconnect

N

Equipment grounding conductor grounds all required equipment...

... and neutral is not bonded to grounding electrode conductor...

G

Equipment grounding conductor

M

...but equipment grounding conductor must be bonded to grounding electrode conductor.

Existing or new grounding electrode

Fig. 5.13 Instead of being an exception, this is now the normally required procedure in Sec. 250-32(a).

feeder just like a normal feeder and treats the subfed building just like any normal load. However, the local electrode provides a local ground reference for the equipment grounding conductors at the remote location.

(2) Grounded systems. This rule applies to situations where buildings are subfed from service equipment located in a separate building with the grounded conductor regrounded, *but only under highly restrictive conditions:*

• No equipment grounding conductor may be run in the supply feeder.

• There must be no continuous metallic paths bonding the grounding systems in both buildings together.

• There must be no ground fault equipment installed at the service of the first building (**Fig. 5.14**). Regrounding a neutral at a second building desensitizes any upstream level GFPE, because fault current in the second building returns over the grounded supply conductor. This means that an arcing fault in the second building would appear to the upstream GFPE sensor as simple line-to-neutral load current

As shown a grounding electrode must be established not only at the main service disconnect, but also at each building or structure subfed from the service. The system grounded (could be a neutral) conductor must be grounded at the supply side of each building or structure disconnect, the disconnect enclosure (if of metal) must be grounded, and a grounding conductor must be installed to the grounding electrode. This basically treats the feeder as if it were a service.

(c) Ungrounded systems. Sec 250-32(c) requires that where the electrical system is ungrounded, the grounding electrode conductor shall only be connected to the subfed building disconnect means

(d) Disconnecting means located in separate building or structure on the same premises. Where the subfed building disconnecting means is/are remote from the subfed building, the grounded circuit must *not* be connected to the grounding electrode system at the subfed building. Where one or more buildings or structures under single management have their disconnecting means located in another building, several rules must all be met. Note that this is only permitted under Sec. 225-32 Ex. 1.

• There must be no connection made at the outbuilding between the grounded circuit conductor and the grounding electrode.

• An equipment grounding conductor for grounding noncurrent-carrying equipment and other items that must be grounded must be ex-

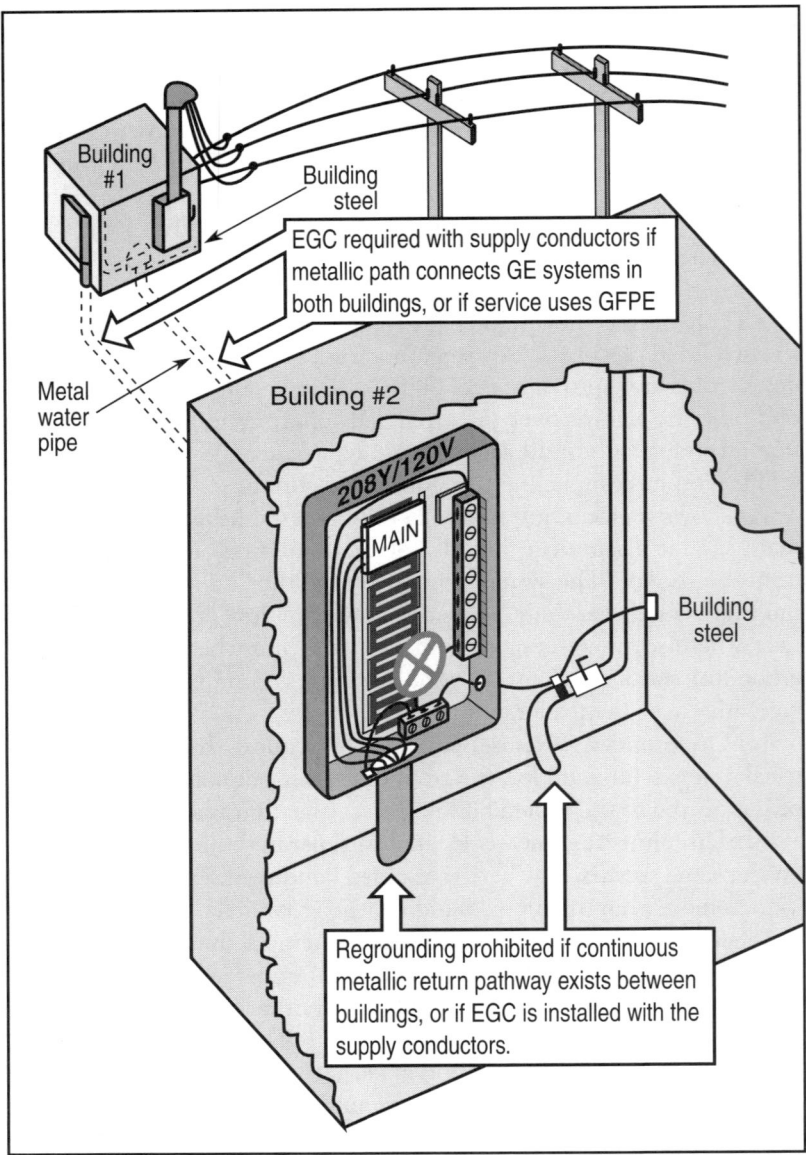

Building
#1

Building
steel

EGC required with supply conductors if
metallic path connects GE systems in
both buildings, or if service uses GFPE

Metal
water
pipe

Building #2

208Y/120V

MAIN

Building
steel

Regrounding prohibited if continuous
metallic return pathway exists between
buildings, or if EGC is installed with the
supply conductors.

Fig. 5.14 Regrounding feeders in second buildings is now severely restricted.

tended to the separate building with the circuit conductors and be connected to an existing grounding electrode system at the separate building.

• If there is more than one circuit within the outbuilding and there
are no existing grounding electrodes there, then a grounding electrode
that complies with Part C of the article must be established and the
equipment grounding conductor must be bonded to it.

• The connection between the grounding electrode and the equipment grounding conductor can be made within a panelboard, within a
junction box as shown in **Fig. 5.15**, or within a similar enclosure located
immediately inside of or outside of the separate building or structure.

(e) Agricultural buildings. Sec. 250-32(e) requires that where live-
stock is housed, that portion of the equipment grounding conductor installed underground to the disconnecting means must be of copper and
must be insulated or covered. An FPN refers to Sec. 547-8 for special
grounding requirements for agricultural buildings.

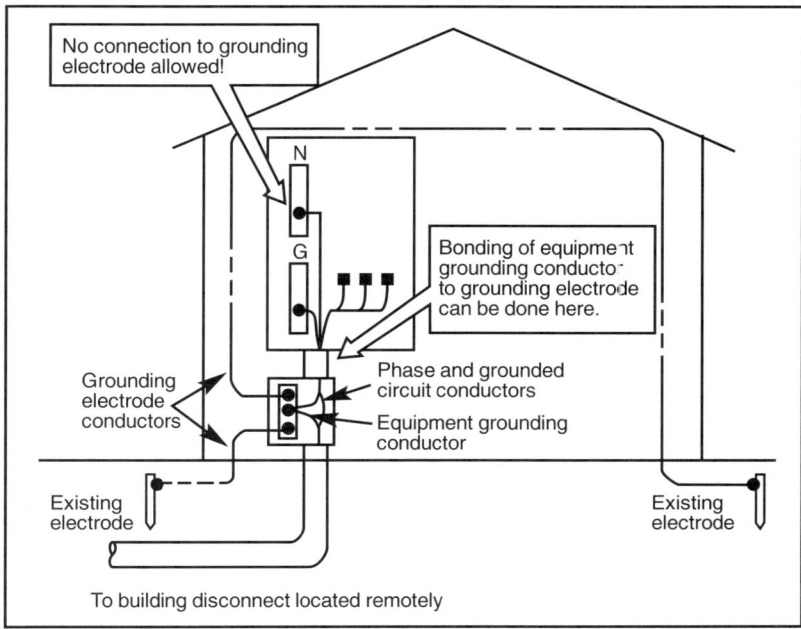

No connection to grounding
electrode allowed!

N

G

Bonding of equipment
grounding conductor
to grounding electrode
can be done here.

Grounding
electrode
conductors

Phase and grounded
circuit conductors

Equipment grounding
conductor

Existing
electrode

Existing
electrode

To building disconnect located remotely

Fig. 5.15 Grounding at outbuilding fed from disconnect located in another building.

Sizing of the grounding conductor to the grounding electrodes is discussed in Chapter 11 of this book.

SEC. 250-36 – HIGH-IMPEDANCE GROUNDED NEUTRAL SYSTEM CONNECTIONS

These systems behave like ungrounded systems in that the first ground fault will not cause an overcurrent device to operate. Instead, alarms required by Sec. 250-36(3) will alert qualified supervisory personnel. The resistance is set such that the current under fault conditions is only slightly higher than the capacitive charging current of the system. Since a fault will often continue until an orderly shutdown can be arranged, the resistor must be continuously rated to safely handle this duty.

This differs from low-impedance grounding systems often used on medium-voltage systems. There, a fault is held to a value that will not cause extensive damage; but will instead result in prompt de-energization of the circuit.

Some additional rules must be observed for high-impedance grounded neutral systems. They are contained in the following subsections.

(a) **Grounding impedance location.** As shown in **Fig. 5.16**, the grounding impedance must be installed between the system neutral and the grounding electrode conductor.

Where a system neutral is not available, the grounding impedance must be installed between the neutral derived from a grounding transformer and the grounding electrode conductor.

(b) **Neutral conductor.** The neutral conductor between the neutral point and the grounding impedance must be fully insulated.

The sizing of the neutral conductor is discussed in Chapter 11 of this book.

(c) **System neutral connection.** Contrary to the Sec. 250-24 rule that requires that the grounded conductor be run to the service disconnect and bonded to its enclosure, when the system is high-impedance grounded, the grounded conductor is prohibited from being connected to ground except through the grounding impedance.

An FPN refers to the latest issue of ANSI/IEEE 142-1991, *Recommended Practice for Grounding of Industrial and Commercial Power Systems* for information on sizing the impedance.

(d) **Neutral conductor routing.** Another difference when a high-impedance grounded system is installed is that the neutral conductor connecting the transformer neutral point to the grounding impedance is not required to be installed with the phase conductors. It can be installed in a separate raceway to the grounding impedance.

The following subsections give the details of how the connections are to be made at the service entrance equipment when a high-impedance grounding system is employed.

(e) **Equipment bonding jumper.** An equipment bonding jumper must be installed unspliced from the first system disconnecting means or overcurrent device to the grounded side of the grounding impedance.

(f) **Grounding electrode conductor location.** The grounding electrode conductor can be attached at any point from the grounded side of

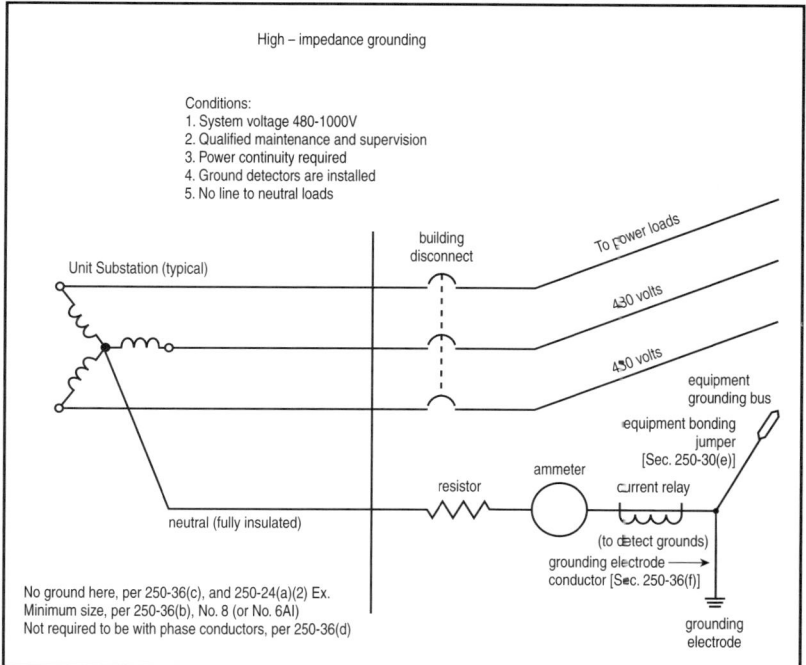

Fig. 5.16 These systems can also use a neutral derived from a zig-zag grounding transformer connected to a delta system.

the grounding impedance to the equipment grounding connection at the service equipment of the first system disconnecting means.

SEC. 250-164-POINT OF CONNECTION FOR DC SYSTEMS

The general rule is that the grounding connection is to be made at one or more supply points, but not at individual services, nor at any point on the premises wiring. This is shown in **Fig. 5.17.**

Sec. 250-164(b), however, states that where the DC system source is located on the premises, a grounding connection is to be made either:

• at the source;

• at the first system disconnecting means or overcurrent device; or

• by some other means that provides the same system protection by using equipment listed and identified for the purpose.

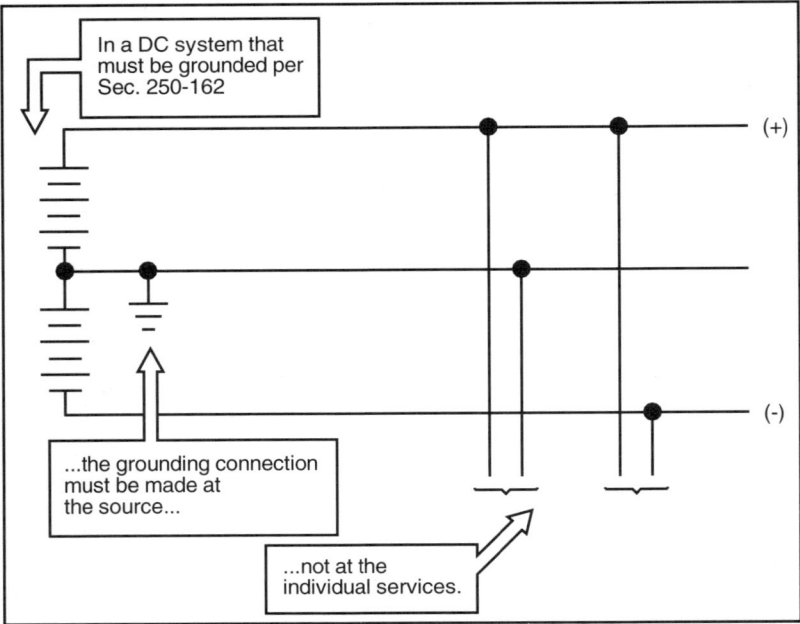

Fig. 5.17 The general rule for grounding DC systems.

ARTICLE 250, PART D — ENCLOSURE, RACEWAY, AND SERVICE CABLE GROUNDING

Prevention of shock hazards that potentially may injure persons contacting electrical equipment is accomplished primarily by following the rules detailed in Parts D and F of NEC Article 250. Both identify the types of items that must be grounded. Part D deals primarily with service raceways or other enclosures containing service conductors, and then adds a general statement in Sec. 250-86 requiring grounding for most other enclosures. Part F, beginning with Sec. 250-110, covers equipment grounding generally. The first four sections in Part F (Sec. 250-110 through 250-116), however, describe categories of equipment that must be grounded, so we are including them in this chapter.

SEC. 250-80–SERVICE RACEWAYS AND ENCLOSURES

Metal enclosures for service conductors and equipment must be grounded. An exception is allowed for situations, such as that shown in **Fig. 6.1,** where a rigid metal conduit elbow has been installed below grade in a run of underground nonmetallic conduit. This is a common technique used to prevent the pull wire from damaging the elbow during installation of the service conductors. To be permitted, however, isolation of the elbow from possible contact is required. No part of the elbow may be closer to grade level than 18 in. (457mm).

SEC. 250-84–UNDERGROUND SERVICE CABLE OR CONDUIT

Where served from a continuous underground metal-sheathed cable system, the sheath or armor of underground service cable metallically

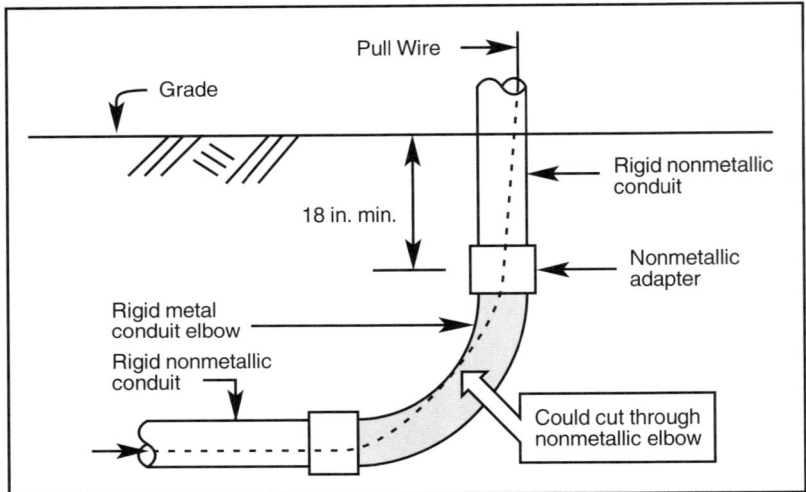

Fig. 6.1. A steel elbow in a run of rigid nonmetallic conduit is not required to be grounded in certain cases.

connected to the underground system, or an underground service conduit containing a metal-sheathed cable bonded to the underground system, is not required to be grounded at the building and is permitted to be insulated from the interior conduit or piping.

This exception from bonding requirements was discussed previously in Chapter 3 of this book.

SEC. 250-86–OTHER CONDUCTOR ENCLOSURES AND RACEWAYS

Metal enclosures that contain conductors other than service conductors must also be grounded. A blanket exemption, however, applies to limited-energy applications relieved from equipment grounding by Sec. 250-112(i); that is in cases where the circuit need not have a system grounding connection in accordance with any applicable provisions in Part B.

Exception No.1 exempts from grounding those metal enclosures for conductors added to existing installations of: open wire; knob-and-tube wiring; and nonmetallic-sheathed cable.

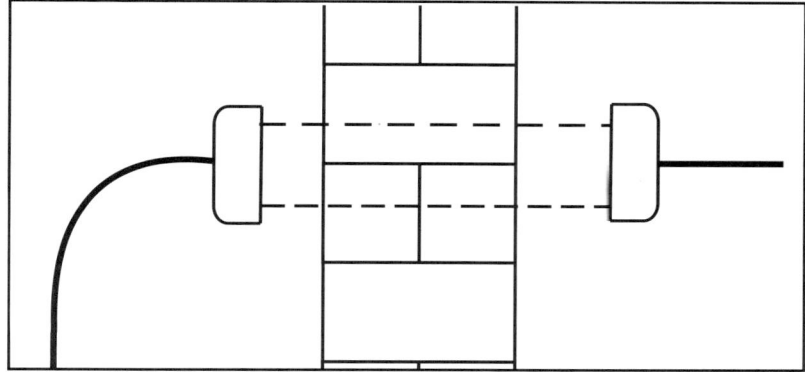

Fig. 6.2. A short run of metal conduit used to protect a cable assembly from damage is not required to be grounded.

There are, however, several qualifiers to this exemption, all of which must be met. They are:

- The existing system contains no equipment ground;
- The run length is less than 25 ft. (7.62 m);
- The enclosure is free from probable contact with ground, grounded metal, metal lath, or other conductive material; and
- The enclosure is guarded against contact by persons.

Exception No.2 says that grounding is not required for short sections of metal enclosures used to support or protect cable assemblies from physical damage (see **Fig 6.2**).

Exception No.3 is a corollary to the exception to Sec. 250-80 that exempts a metal elbow in a nonmetallic conduit run from being grounded. The same conditions apply, including the 18-in. burial depth requirement.

SEC. 250-110–EQUIPMENT FASTENED IN PLACE OR CONNECTED TO PERMANENT WIRING METHODS (FIXED)

Exposed noncurrent-carrying metal parts of fixed equipment likely to become energized must be grounded when they meet the conditions specified in the following subsections.

Note that fixed equipment is equated to equipment that is either fastened in place *or* connected with a permanent wiring method — either condition, if true, creates fixed equipment subject to these rules. Sec. 370-28(c) also requires metal box covers to be grounded in accordance with these rules. In addition to conventional boxes, watch out for this on such items as handhole covers in public sidewalks and the like.

(1) Vertical and horizontal distances. Where within 8 ft. (2.44 m) vertically or 5 ft. (1.52 m) horizontally of ground or grounded metal objects and can possibly be contacted by persons. These dimensions generally place the metal parts within reach.

(2) Wet or damp locations. Where the equipment is in such a location and is not isolated. Article 100 defines isolated as: Not readily accessible to persons unless special means of access are used.

(3) Electrical contact. Where the equipment is in electrical contact with metal.

(4) Hazardous (classified locations). Where the equipment is located in a hazardous area defined in Articles 500 through 517.

(5) Wiring methods. Where supplied by a metal clad, metal-sheathed cable, metal raceway, or other wiring method that provides an equipment ground. Short sections of metal enclosures used for support or protection (as noted in Ex. 2 to Sec 250-86) are not included as part of this requirement.

(6) Over 150V to ground. Where equipment operates with any terminal at over 150V to ground. There are, however, a few exceptions which apply to all of these rules.

Exception No. 1 relieves metal frames of electrically heated appliances from being grounded where they have been exempted by special permission. They must, however, be permanently and effectively insulated from ground.

Exception No. 2 excludes from the grounding requirement the cases of transformers, capacitors, and other distribution equipment mounted on wood poles at a height of at least 8 ft. (2.44 m) above ground or grade level (see **Fig. 6.3**). This includes such equipment operating over 1000V to ground.

Exception No. 3 excludes listed equipment protected by a system of double insulation, or its equivalent, from being grounded. Such equipment must be distinctively marked so it can be easily identified.

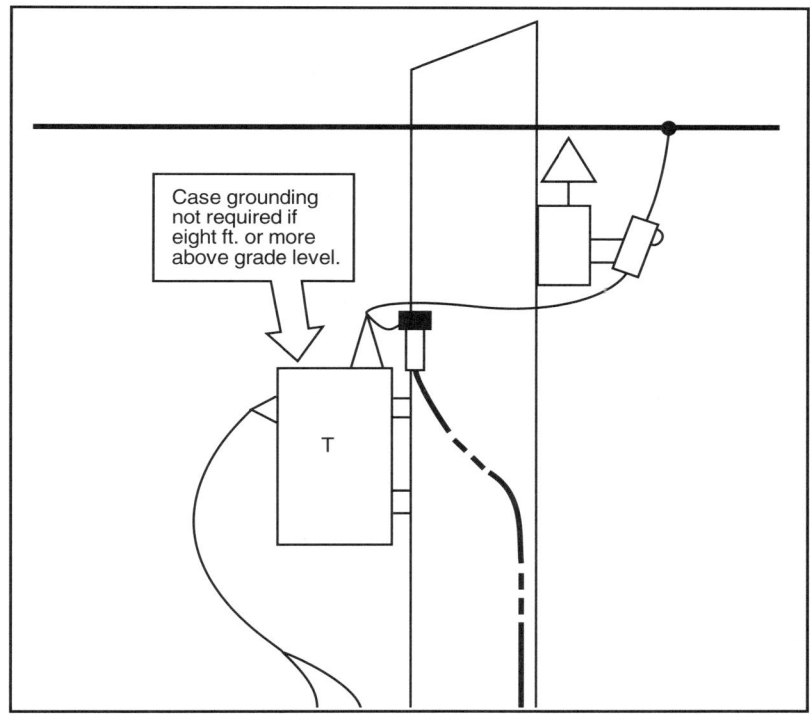

Fig. 6.3. Distribution transformers on wood poles are mounted over 8 ft. from the ground, making it unnecessary to ground their cases.

SEC. 250-112–FIXED EQUIPMENT FASTENED IN PLACE OR CONNECTED BY PERMANENT WIRING METHODS

Exposed noncurrent-carrying parts of the equipment, enclosures, and facilities described in the following subsections must be grounded whenever they are exposed, regardless of voltage.

(a) **Switchboards.** The frames and supporting structures. An exception is made for 2-wire DC switchboards that are effectively insulated from ground.

(b) Pipe organs. Generator and motor frames of an electrically operated unit, unless effectively insulated from ground and the motor driving it..

(c) Motors. Frames of stationary motors where:

• They are supplied by metal-enclosed wiring;

• They are in a wet location and are not isolated or guarded;

• They are in a hazardous location covered in Articles 500 – 517; or

• The motor operates with any terminal at over 150V to ground.

Note that in any case where a motor is ungrounded, it must be permanently and effectively insulated from the ground.

(d) Motor controllers. Enclosures, except those attached to ungrounded portable equipment, and the lined covers of snap switches.

(e) Elevators and cranes. Electrical equipment on the units.

(f) Garages, theaters, and motion picture studios. Electrical equipment for these locations. An exception is made for pendant lampholders supplied by circuits of not over 150V to ground.

(g) Electric signs. Frames and associated equipment. Sec. 600-7 includes more detailed requirements, including:

• Up to 100 ft of greenfield or liquidtight flexible metal conduit can be used in a bonding path to sign elements on the load (secondary) side of a neon sign transformer or power supply. The currents involved are very low and the 100 ft distance was investigated by a testing laboratory and found suitable. This is a Chapter 6 amendment to Sec. 250-118(6)(c).

• Small metal parts not exceeding 2 in. in any dimension and spaced at least ¾-in from neon tubing don't require a bonding connection. This removes such incidental equipment as the metal feet on tubing support clips.

• Bonding conductors must be copper and at least #14, this rule having been modified and relocated from its former location in this article.

• If a nonmetallic wiring method encloses secondary wiring from a neon transformer or power supply, the bonding conductor must run outside the conduit spaced at least 1½-in. distant for power supplies running below 100 hz, and 1¾-in. away for higher frequencies.

This provision, unique in the history of the NEC, has been well substantiated. Separating the grounding conductor reduces the intensity of the voltage gradient, which in turn reduces corona. There have been fires produced from the deterioration of the GTO cable from corona discharges. Obviously this separation isn't possible in a metal raceway. The metal raceway distributes the ground plane equally around the conductor, and in addition, Sec. 600-32(j) more severely limits the allowable length on metal raceways to the first neon terminal (to 20 ft instead of 50 ft). Most of the voltage problem occurs in this particular segment of the secondary wiring system.

(h) **Motion picture projection equipment.** All such equipment.

(i) **Power-limited Remote-control, signaling, and fire-protective signaling circuits.** Equipment supplied by Class 1, Class 2, and Class 3 circuits where system grounding is required in Part B of Article 250. For example, an exposed metal solenoid on a 24V Class 2 control circuit derived from a 277V to 24V transformer must be equipment grounded because Sec. 250-20 requires the secondary to be grounded. This is often overlooked. (see **Fig. 6.4**)

(j) **Lighting fixtures.** Units with exposed metal parts must be grounded unless they are insulated from ground and other conducting surfaces, or accessible only to qualified personnel. If a fixture is directly wired or

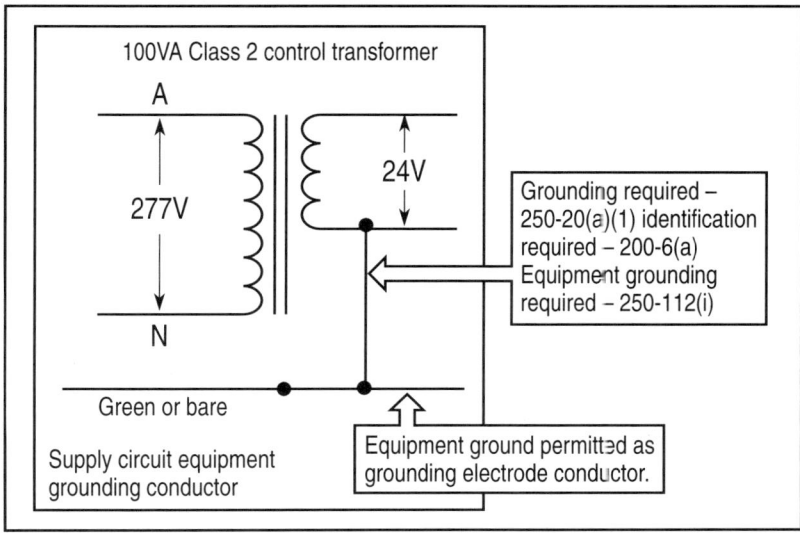

Fig. 6.4. The colors effectively reserved by this rule (white and green) may not agree with the control circuit diagram provided with the equipment.

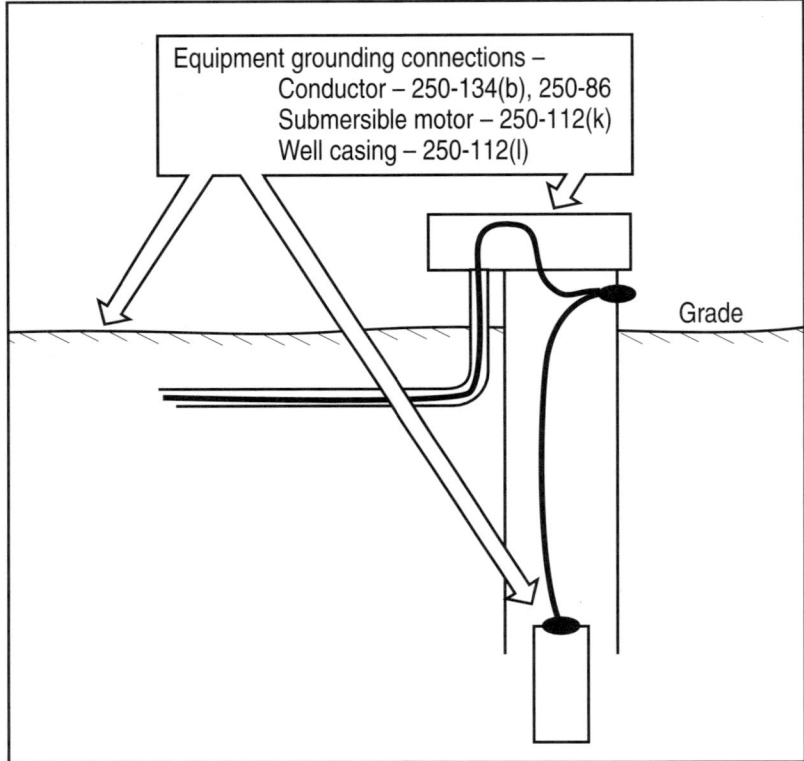

Equipment grounding connections –
Conductor – 250-134(b), 250-86
Submersible motor – 250-112(k)
Well casing – 250-112(l)

Grade

Fig. 6.5 Well casings are potential grounding electrodes per Sec. 250-52(b), but even if so used, an equipment grounding conductor must be run with the circuit conductors.

attached to an outlet supplied by a wiring method that does not provide an equipment grounding conductor, then the fixture must be made of insulating material, and have no exposed conductive parts.

(k) **Skid mounted equipment and skids** must be grounded.

(l) **Motor-operated water pumps.** All units including the submersible types. *Note that this item and the next apply whether or not they are exposed.* (see **Fig. 6.5**)

(m) **Metal well casings.** Where a submersible pump is used in a metal well casing, the well casing must be bonded to the pump circuit equipment grounding conductor.

The exposed noncurrent carrying parts of motor-driven water pumps are among the items that are specifically required to be grounded.

SEC. 250-116–NONELECTRIC EQUIPMENT

Structural metal is widely used in buildings, supports, decorations, and the like. Although not a direct part of a piece of electrical equipment, some are required by the NEC to be grounded. The items requiring grounding are listed in the following subsections.

(1) **Cranes.** Frames and tracks of electrically operated cranes.

(2) **Elevator cars.** Frames of nonelectrically driven elevator cars to which electrical conductors are attached.

(3) **Electric elevators.** Hand-operated metal shifting ropes or cables of electric elevators.

It should also be noted that for mobile homes and RV's, grounding is required per Articles 550 and 551 as follows.

Mobile homes. The following items are required to be grounded [per Sec. 550-11(b) and (c)].

- Grounding bus in the distribution panelboard or disconnecting means (by a green-colored insulated ground wire in the supply cord or permanent wiring).
- All exposed metal parts, enclosures, frames, lamp fixture canopies, etc. (by bonding to the grounding terminal of the distribution panel).
- Cord-connected appliances, such as washing machines, dryers, refrigerators, the electrical system of gas ranges, etc. (by means of a cord with a grounding conductor and grounding-type attachment plug).
- All exposed noncurrent-carrying metal parts that may become energized (by a bonding conductor connected to the grounding terminal in the distribution panelboard).
- The chassis (by a bonding jumper from the distribution panelboard to an accessible terminal on the chassis).
- Metallic gas, water , and waste pipes and metallic air circulating ducts (considered bonded if connected to the terminal on the chassis by clamps, solderless connectors, or grounding-type strap).

The metallic roof and exterior covering of a mobile home is considered to be bonded if:

- The metal panels overlap one another and are securely attached to the wood or metal frame parts by metallic fasteners; and
- The lower panel of the metallic exterior covering is secured by metallic fasteners at a cross member of the chassis by two metal straps (at opposite ends) per mobile home unit or section.

Recreational vehicles. With some minor differences, the requirements are the same as for mobile homes. An exception to the requirement for chassis grounding via a bonding jumper is made for RV's with a unitized metal chassis-frame construction to which the distribution panelboard is securely fastened by bolts, nuts, welding, or riveting. Furnace and metal circulating air ducts also must be bonded.

SEC. 250-114–EQUIPMENT CONNECTED BY CORD AND PLUG

Exposed noncurrent-carrying metal parts of cord-and-plug connected equipment that is likely to become energized must be grounded. These subsections define the types of equipment involved in the following oc-

Cord-and-plug hand tools are required to be grounded unless they are protected by a system of double insulation.

cupancies. There is a broad exception for double-insulated equipment, however.

Listed double-insulated tools, appliances and other equipment in the following list are not required to be grounded, but they must be distinctly marked so they can be readily identified as being double-insulated. This exception does not apply to hazardous (classified) locations.

(1) In hazardous (classified) locations. The general requirements for grounding are given in Chapter 5 and must be followed in any locations meeting the defined class hazard. Refer to the end of this chapter for this information.

(2) Where operated at over 150 volts to ground, except that guarded motors and insulated frames of electrically-heated appliances are not required to be grounded. Note that the motor allowance is qualified by Sec. 430-142, which requires that in any case where a motor is ungrounded, it must be permanently and effectively insulated from the ground, not just guarded. In the case of electrically heated appliances, the exception is the same as for permanent connections in Sec. 250-110 Ex. 1, and special permission is needed.

(3) In residential occupancies. Grounding is required for the following items:

• Refrigerators, freezers, and air conditioners;

• Clothes-washing, clothes-drying, kitchen waste disposers, dish-washing machines, information technology equipment, sump pumps, and electrical aquarium equipment;

• Hand-held motor-operated tools, stationary and fixed motor-operated tools, and light industrial motor-operated tools;

• Motor-operated appliances including hedge clippers, lawn mowers, snow blowers, and wet scrubbers; and

• Portable handlamps

(4) In other than residential occupancies. The following items must be grounded:

• Refrigerators, freezers, and air conditioners;

• Clothes-washing, clothes-drying, and dishwashing machines;

• Information technology equipment, sump pumps, and electrical aquarium equipment;

• Hand-held motor-operated tools, stationary and fixed motor-operated tools, and light industrial motor-operated tools;

• Motor-operated appliances including hedge clippers, lawn mowers, snow blowers, and wet scrubbers;

• Cord-and-plug connected appliances used in damp or wet locations or by persons standing on the ground or on metal floors or working inside metal tanks or boilers;

• Tools likely to be used in wet or conductive locations; and

• Portable handlamps.

There is an exception to these requirements, applicable to tools and handlamps likely to be used in wet or conductive locations. They are not required to be grounded if supplied through an isolating transformer with an ungrounded secondary of not over 50V.

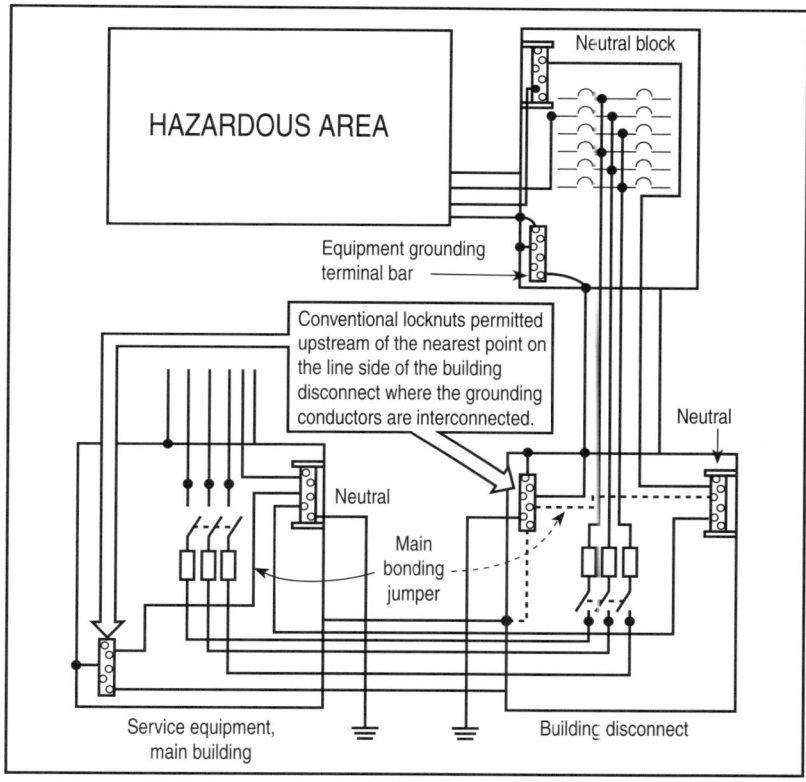

Fig. 6.6 Given the provisions of Sec. 250-32(b)(2), the bounding jumper on the building disconnect probably would not be installed. Enhanced grounding would then be required all the way back to the service equipment.

HAZARDOUS (CLASSIFIED) LOCATIONS

The main difference with respect to grounding in hazardous locations is that enhanced equipment grounding requirements apply, not just within the hazardous location, but also in the entire home run of circuits supplying such locations. The rule applies to the source of a separately derived system, or to upstream disconnects until a bonding jumper connects equipment grounding and grounded circuit conductors. See **Fig. 6.6.**

• Class I locations. The grounding requirements in these areas are defined in Sec. 501-16.

• Class II locations. The grounding requirements in these areas are defined in Sec. 502-16.

• Class III locations. The grounding requirements in these areas are defined in Sec. 503-16.

• Zone-classified areas follow Class I rules, because Sec. 505-25 makes compliance with Sec. 501-16 mandatory for these areas.

In addition to these requirements, which refer back to requirements in Article 250, there are these specific items listed in Secs. 504 through 516 that must considered when providing grounding.

Intrinsically safe systems (per Secs. 504-50 and 504-60). Apparatus, associated apparatus, cable shields, enclosures, and raceways (if of metal) must be grounded. An FPN refers to the latest issue of ANSI/ISA RP 12.6 Installation of Intrinsically Safe Instrument Systems in Class I Hazardous Locations for more details on supplementary bonding that may be needed.

Commercial garages, repair and storage (per Art. 511). The grounding requirements for the class of hazard involved must be met, including Sec. 511-6(c) on grounding continuity.

Aircraft hangers. Besides the general grounding requirements for the

Hazardous locations, including aircraft hangers, are required to meet the grounding requirements of Article 250.

class of location involved:

- Sec 513-5 requires the flexible cord for pendants and portable utilization equipment within a Class I location to contain a separate equipment grounding conductor, and there are enhanced grounding continuity provisions;
- Sec. 513-11(c) (aircraft energizers and ground support equipment) and Sec. 513-12(b) (mobile servicing equipment) also have the requirement that their flexible power cords include an equipment grounding conductor; and
- Sec. 513-16 says that all metal raceways, metal-jacketed cables, and all noncurrent-carrying metal portions of fixed or portable equipment, regardless of voltage, must be grounded.

Gasoline dispensing service stations. Sec. 514-16 requires that metal portions of dispensing pumps, metal raceways, metal-jacketed cables, and all noncurrent-carrying metal parts of electric equipment, regardless of voltage, be grounded.

Bulk storage plants. Sec. 515-16 requires that all metal raceways, metal-jacketed cables, and all noncurrent carrying metal parts of electric equipment be grounded. An FPN refers to Secs. 6-4.4.1.2 and 5-4.4.1.7 of the latest issue of NFPA 30, Flammable and Combustible Liquids Code for more information on grounding for static protection.

Spray applications, dipping, and coating processes. The following specialized requirements are listed.

- Sec. 516-3(d) Exception No. 2 permits portable electric lamps and other utilization equipment to be used in a spray area if, among other requirements, all metallic parts of the drying apparatus are electrically bonded and grounded.
- Sec. 516-4(f) requires that all electrically conductive objects that are part of fixed electrostatic equipment and are used in the spray area (except those objects required by the process to be at high voltage) must be grounded. The equipment subject to this provision includes paint containers, wash cans, guards, hose connectors, brackets, and any other electrically conductive objects or devices in the area.
- Sec 516-5(c) requires that the handle of an electrostatic hand-spray gun be electrically connected to ground to prevent a buildup of static electricity.

- Sec. 516-5(d) requires paint containers, wash cans, and other electrically conductive objects used in electrostatic hand spraying to be adequately grounded.
- Sec 516-5(e) states that objects to be painted via electrostatic hand spraying must be maintained in metallic contact with the conveyor or other grounded supports.
- Sec 516-6(a) requires that portable electric lamps or other utilization equipment that are permitted to be used during cleaning and repair operations in a powder coating process must all have metal parts effectively grounded.
- Sec. 516-6(d) also requires that transformers, power packs, control apparatus, and other electric portions of equipment located outside of an electrostatic fluidized bed be effectively grounded. In addition, all electrically conductive objects within the powder-coating area must be adequately grounded. Also, all objects being coated must be maintained in electrical contact with the conveyor or other support to ensure proper grounding.
- Sec. 516-16 requires the grounding of all metal raceways, metal-jacketed cables, and all noncurrent-carrying metal parts of fixed or portable equipment, regardless of voltage.

7

ARTICLE 250, PART G — METHODS OF EQUIPMENT GROUNDING

Not only is it necessary to understand what equipment and systems are to be grounded, but also the methods for achieving grounding that will satisfy the Code rules. The applicable requirements are spelled out in Part G of the NEC. In addition to the material in Part G, there are three sections in Part A (General) that directly apply to these concepts, so we are including them here.

Exothermic welding provides a reliable connection cable-to-cable, cable-to-rod, cable-to-flat surface, and others.

SEC. 250-8–GROUNDING ELECTRODE CONNECTIONS TO CONDUCTORS AND EQUIPMENT

Grounding electrode conductors and bonding jumpers must be connected by:
- Exothermic welding;
- Listed pressure connectors;
- Listed clamps; or
- Other listed means.

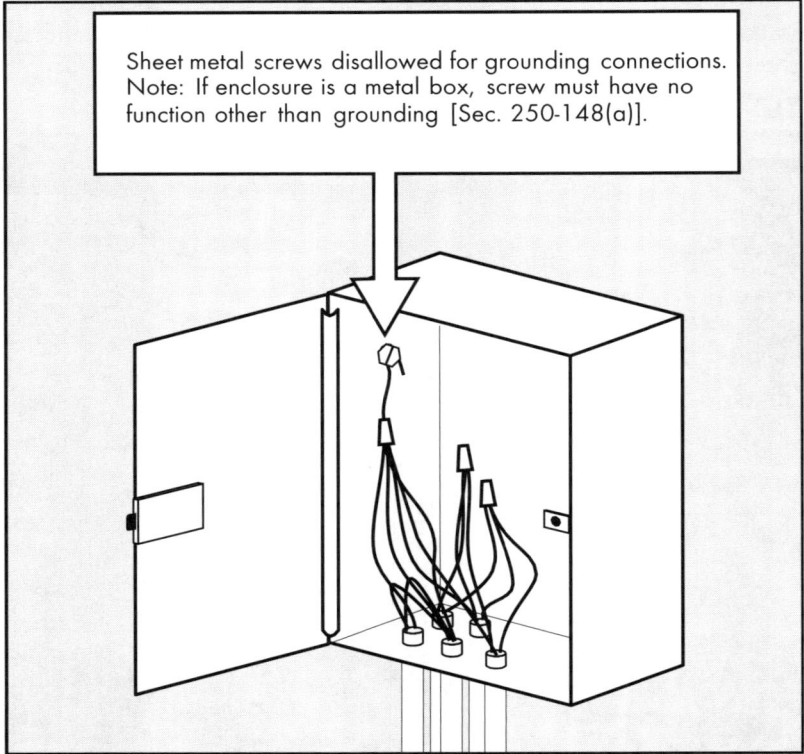

Sheet metal screws disallowed for grounding connections. Note: If enclosure is a metal box, screw must have no function other than grounding [Sec. 250-148(a)].

Fig. 7.1 Don't use sheet metal screws for grounding terminations. Self tapping screws with machine threads are OK.

Connection devices that depend solely on solder are not to be used for the purpose. In addition, sheet metal screws can't be used to connect a grounding conductor to an enclosure. Sheet-metal screws have a very coarse pitch, which means their holding power in metal enclosures isn't very high due to the low mechanical advantage of the thread. They frequently loosen up over time. Although they work fairly well in wood (and indeed, the new rule doesn't rule out wood screws), the only time that would be useful would be if wood were behind the metal enclosure, helping to hold the screw. That would mean the screw would no longer be only functioning for grounding, a violation, at least for metal boxes, of Sec. 250-148(a). See **Fig. 7.1.**

SEC. 250-10 – PROTECTION OF ATTACHMENT

Ground clamps or other fittings must be approved for the general use without protection, or they must be protected from physical damage as

Serrated connecting surfaces on this conduit grounding fitting will bite through nonconductive coatings of boxes or enclosures.

indicated in the following subsections.

Not likely to be damaged. The fittings can be installed without extra protection from physical damage in locations where they are not likely to be damaged.

Protective covering. The fittings can also be protected by being enclosed in metal, wood, or equivalent protective coverings.

SEC. 250-12–CLEAN SURFACES

Nonconductive coating such as paint, lacquer, and enamel on equipment to be grounded must be removed from threads and other contact surfaces to assure good electrical continuity or be connected by types of fittings so designed as to make such removal unnecessary. This includes such items as locknuts with sharp teeth and hubs with serrated contact surfaces.

SEC. 250-130–EQUIPMENT GROUNDING CONDUCTOR CONNECTIONS

The details of how equipment grounding conductor connections at service equipment are to be made are listed in the following subsections.

(a) For grounded system. Systems energized from a single-phase 3-wire, or 3-phase 4-wire source are generally grounded systems. As shown in **Fig 7.2**, the connection is to be made by bonding the equipment grounding conductor to the grounded service conductor and grounding electrode conductor. For a separately derived system, the procedure in Sec. 250-30(a)(1) is exactly the same, except instead of a grounded service conductor, there is a grounded system conductor.

(b) For ungrounded system. In 3-phase, 3-wire systems, there is no neutral to bring into the service equipment. In this case, the connection is made by bonding the equipment grounding conductor to the grounding electrode conductor. For a separately derived system, the procedure in Sec. 250-30(b)(1) is exactly the same. The connection is made at any point between the source and the first system disconnecting means.

(c) A special rule applies for **nongrounding receptacle replacements or branch circuit extensions.** We cover these cases at the end of this chapter.

SEC. 250-132 – SHORT SECTIONS OF RACEWAY

Isolated sections of metal raceway or cable armor, where required to be grounded, must be grounded in accordance with Sec. 250-134. The most notable case where grounding is mandated is described in Sec. 250-92(a)(3) and Sec. 250-64(d) where the raceway encloses a grounding electrode conductor. This requirement is discussed in Chapter 2 of this book.

SEC. 250-134 – EQUIPMENT FASTENED IN PLACE OR CONNECTED BY PERMANENT WIRING METHODS (FIXED) - GROUNDING

Noncurrent-carrying metal parts of equipment, raceways, and other enclosures, if grounded, must be grounded by one of the methods indicated in the following subsections.

This requirement used to say "where required to be grounded." The new wording reflects an important philosophical point. The point is that

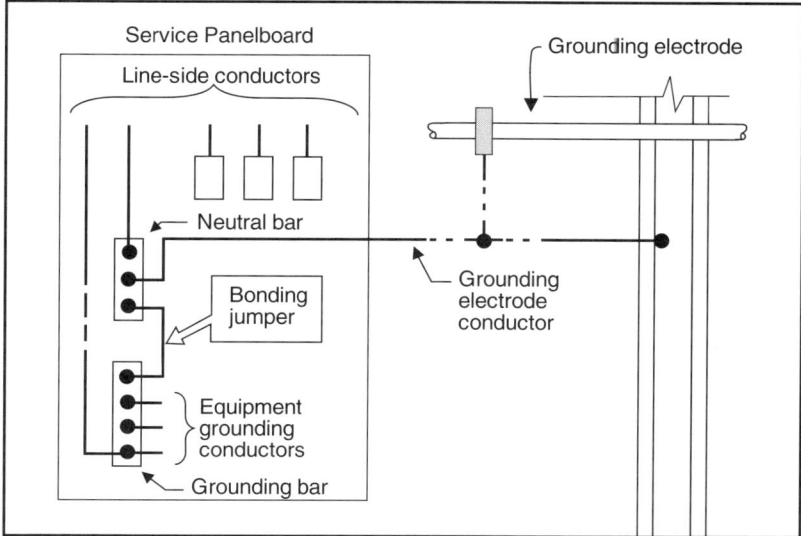

Fig. 7.2 Typical equipment grounding conductor connection at a grounded-system service entrance.

once you undertake to provide grounding, do it correctly, because others will assume it so and proceed accordingly.

An exception to this requirement is made for items that are permitted to be grounded by connection to the grounded circuit conductor (neutral). This is principally limited to instances where a neutral is re-grounded in a second building, as covered in Sec. 250-32(b), and also to particular equipment installations covered in Sec. 250-140 and Sec. 250-142, which are covered later in this chapter.

(a) Equipment grounding conductor types. Equipment can be grounded by any of the grounding conductors listed in Sec. 250-118. These include:

- Copper or other corrosion-resistant conductor;
- Rigid metal conduit;
- IMC;
- EMT;
- Flexible metal conduit where both the conduit and fittings are listed for grounding
- Listed flexible metal conduit that is not listed for grounding but meeting certain conditions
- Listed liquidtight flexible metal conduit that is not listed for grounding but meeting certain conditions
- Flexible metallic tubing where the tubing is terminated in fittings listed for grounding and meeting certain conditions
- Armor of Type AC cable;
- The copper sheath of Type MI cable;
- The metallic sheath or combined metallic sheath and grounding conductors of Type MC cable;
- Cable trays, as permitted by Sec. 318-3(c) and 318-7;
- Cablebus, as permitted by Sec. 365-2(a); and
- Other continuous metal raceways listed for grounding.

The requirements for each item used for this purpose are discussed in Chapter 8 of this book.

(b) With circuit conductors. Equipment can also be grounded by a grounding conductor contained within the same raceway, cable, or cord, or otherwise run with the circuit conductors.

There are two exceptions. Ex. 1 applies to dc circuits only. The equipment grounding conductor is permitted to be run separately from the circuit conductors.

Ex. 2 exempts the grounding conductor from having to be run with the circuit conductors when it involves a grounding-type receptacle that has replaced a nongrounding type receptacle. This is discussed further in Sec. 250-130(c) at the end of this chapter.

Basic Principle: Both these approaches, (a) and (b), have one thing in common: *the equipment grounding path runs with or encloses the circuit conductors.* This minimizes the impedance that results when the conductor supplying a fault and the conductor returning power from a fault are separated.

There are many electrical rooms designed with beautiful grounding busbars encircling the room. The conduits go down into the slab from switchboard to switchboard, and each switchboard has an impressive bonding conductor running from frame to bus, and no equipment grounding conductor down in the slab with the circuit conductors.

Attractive though it may be, the pretty grounding bus running around that electrical room is accomplishing little. Fault return current has to run around the outside of the room to get back to its source. The separation can easily double the impedance. Although 0.01 ohm going to 0.02 ohm may not seem like much, at 277V to ground it's the difference between 27,700A and 13,850A.

For a 3000A circuit breaker, the difference in these two numbers could be the difference between the breaker seeing the fault as an overload, perhaps to be ridden out for a while on an inverse-time basis, and fault current in the instantaneous tripping range of the breaker. The difference between those two responses has often been the difference between calling the fire department and not.

SEC. 250-136–EQUIPMENT CONSIDERED EFFECTIVELY GROUNDED

Noncurrent-carrying metal parts of the equipment are considered to be effectively grounded if the conditions listed in the following subsections are met.

(a) Equipment secured to grounded metal supports. Electrical equip-

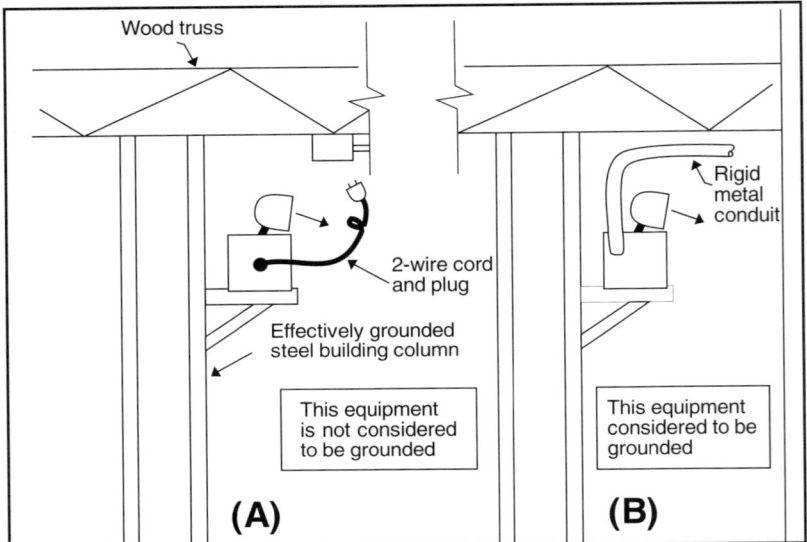

Fig. 7.3 Connecting a piece of electrical equipment to a grounded steel support member does not provide the grounding required by the NEC (A). A metal raceway or internal equipment grounding conductor is needed (B).

ment in electrical contact with, and secured to, a metal rack or structure provided for its support is considered to be grounded if it is grounded by one of the means indicated in Sec. 250-134. As shown in **Fig. 7.3**, the structural metal frame of a building is not to be used as the required equipment grounding conductor for AC equipment.

The reasoning behind this requirement is that the equipment grounding conductor must be run with, or in close proximity to the circuit conductors to limit the impedance to the flow of fault current in AC circuits. Because the same phenomenon does not occur in DC circuits, the equipment grounding conductor is not required to be run with the circuit conductors of such a system. [See permission granted in Sec. 250-134(b) Ex 2.]

(b) Metal car frames supported by metal hoisting cables attached to, or running over, metal sheaves or drums of elevator machines are considered to be grounded – if grounded by one of the methods indicated in Sec. 250-134.

SEC. 250-138–CORD AND PLUG CONNECTED EQUIPMENT

Noncurrent carrying metal parts of this type of equipment that are required to be grounded must be grounded by one of the methods described in the following subsections.

(a) By means of grounding conductor. The equipment grounding conductor must be run with the power supply conductors in a cable assembly or flexible cord properly terminated in grounding-type attachment plug with one fixed grounding contact. The equipment grounding conductor must comply with Sec. 400-23, which means it will be insulated with a green color or braid, or green with yellow stripes.

There is an exception that allows for a movable, self-restoring ground pin, but only for a plug with integral GFCI protection, and operating not over 150V between conductors or to ground. See **Fig. 7.4.**

(b) By means of a separate flexible wire or strap. Where part of the equipment, a separate flexible insulated or bare wire or strap is permitted to serve as the equipment grounding conductor. This wire or strap must be protected from physical damage – "as well as is practicable."

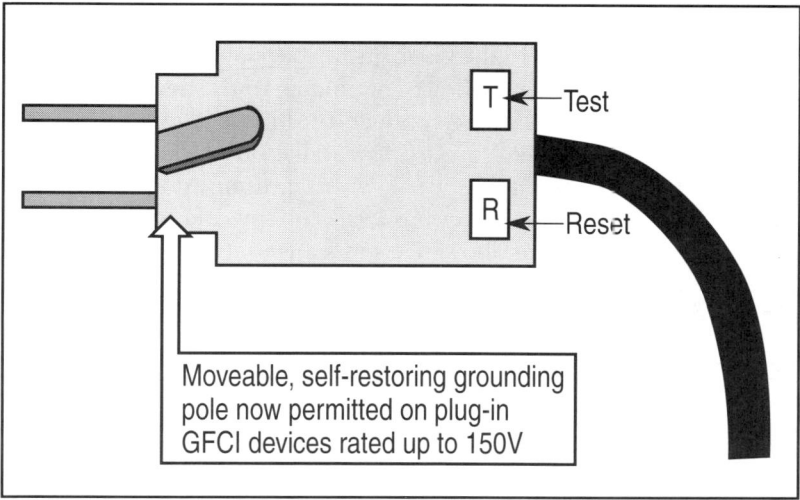

Test

Reset

Moveable, self-restoring grounding pole now permitted on plug-in GFCI devices rated up to 150V

Fig. 7.4 This concept is also covered in Sec. 410-58(a).

Sec. 250-140– Frames of ranges and clothes dryers in new installations must be grounded through the equipment grounding conductor of the power circuit, and this equipment grounding conductor must either be connected directly to the equipment or through a flexible cord with plug and receptacle. However, for existing installations only, the grounded (neutral) conductor can still be used to ground these frames where the following provisions are met:

(1) The installation is not at a mobile home, recreational vehicle, or, per Sec. 553-9, at a floating building.

(2) The supply circuit is 120/240-volt, single-phase, 3-wire; or 208Y/120-volt derived from a 3-phase, 4-wire wye-connected system.

(3) The grounded conductor is not smaller than No. 10 copper or No. 8 aluminum.

(4) The grounded conductor is insulated, or the grounded conductor is uninsulated and part of a Type SE service-entrance cable and the branch circuit originates at the service equipment. This is Type SE cable (Art. 338) only, not Type NM or others, for which the grounded conductor must be insulated, and not used for equipment grounding.

(5) Grounding contacts of receptacles furnished as part of the equipment are bonded to the equipment.

For over 50 years until the 1996 NEC, a "temporary" wartime exception to the general rules in the Code, justified to save copper for the war effort, refused to go away. It allowed grounding to the neutral without a separate equipment grounding conductor for most ranges and clothes dryers. Now it only applies to existing installations. In fairness to those opposed to the change, there was very little, if any, documented loss experience from 3-wire connections. Still, in terms of the technical coherence of the Code, this allowance was an aberration.

SEC. 250-142–USE OF GROUNDED CIRCUIT CONDUCTOR FOR GROUNDING EQUIPMENT

Some equipment types other than those listed in Sec. 250-142 are allowed to be grounded by connection to the grounded (neutral) circuit conductor. These cases, as well as those where such a practice is prohibited, are spelled out in the following subsections.

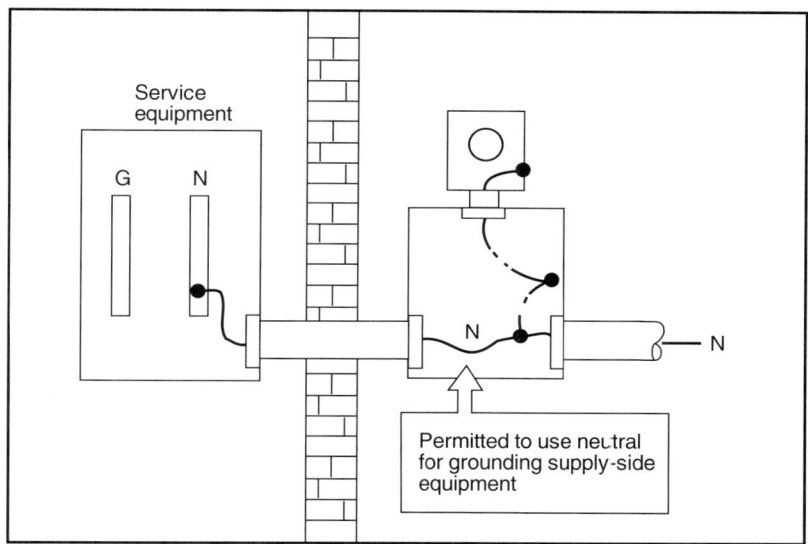

Fig. 7.5 The grounded (could be a neutral) conductor is permitted to be used for grounding equipment on the supply side of the service.

(a) Supply-side equipment such as noncurrent-carrying metal parts of equipment, raceways, and other enclosures are permitted to be grounded by connection to the neutral if they are located:

• On the supply side of, or within, the service disconnecting means for ac systems(see **Fig. 7.5**). Sec. 250-164 makes this a violation for dc systems.

• On the supply side of the main disconnecting means for separate buildings as provided in Sec. 250-32(b); or

• On the supply side of the disconnecting means or overcurrent devices of a separately derived system, as permitted by Sec. 250-30(a)(1).

(b) Load-side equipment. A grounded circuit conductor is *prohibited* from being used for grounding noncurrent-carrying metal parts of equipment on the load side of the service disconnecting means. It is also prohibited to use the grounded circuit conductor as an equipment grounding conductor on the load side of a separately derived system disconnecting means or on the load side of the overcurrent devices for a separately derived system not having a main disconnecting means. There are many exceptions to this general rule.

Exception No. 1 coordinates with Sec 250-140, where permission was granted to connect the neutral to the frames of clothes dryers and ranges in existing installations only.

Exception No. 2 permits meter enclosures on the load side of the service disconnect by connecting to the neutral if:

• No service ground-fault protection is installed, and

• All meter enclosures are located near the service disconnecting means.

• The size of the grounded circuit conductor is not smaller than the size specified in Table 250-122 for equipment grounding conductors.

Exception No. 3 refers to Sec. 250-164 (Chapter 5 in this book), permitting dc systems to be grounded on the load side of the disconnecting means or overcurrent device under limited circumstances.

Secs. 490-72(e)(1) and 490-74 require that all exposed noncurrent-carrying metal parts of medium voltage electrode-type boilers and asso-

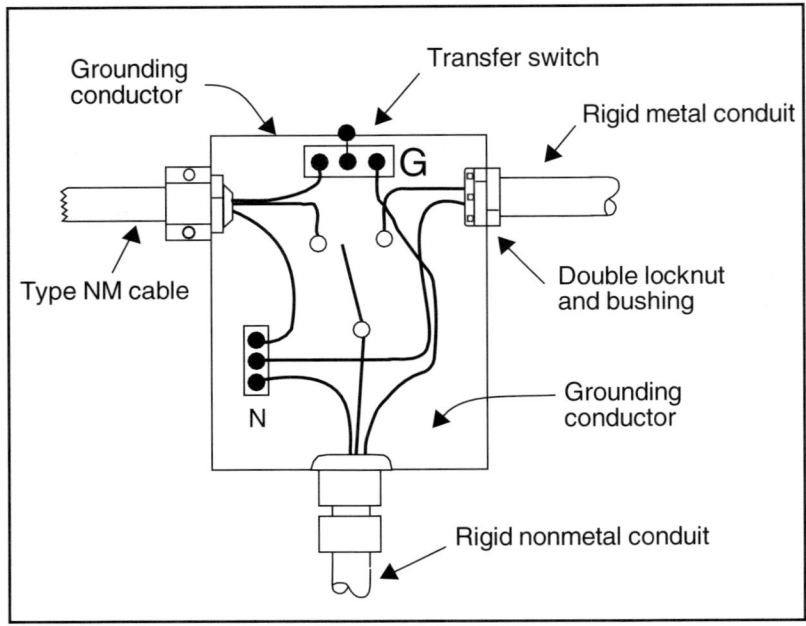

Fig. 7.6 Typical grounding requirements at a transfer switch.

ciated exposed grounded structures or equipment are to be bonded to the pressure vessel or to the neutral conductor to which the vessel is connected. The neutral must be connected to the vessel.

The rules of Sec. 250-102 (Chapter 3 in this book) for equipment bonding jumpers apply, except the ampacity of the bonding jumper must not be less than the ampacity of the neutral conductor. If the medium voltage distribution system doesn't have a neutral, then one must be derived with a grounding transformer to install this equipment.

These rules are in direct conflict with the rules in Art. 250. They used to be located in Chapter 7 where they were protected by Sec. 90-3, but now, relocated into Chapter 4 for the 1999 NEC, you will need permission to apply them. They have been part of the Code for many cycles, and they should be respected until the correlation problem has been corrected.

SEC. 250-144–MULTIPLE CIRCUIT CONNECTIONS

Where equipment is required to be grounded, and is supplied by separate connection to more than one circuit or grounded premises wiring system, a means of grounding must be provided at each such connection as specified in Secs. 250-134 and 250-138. **Fig 7.6** shows a typical example of where this provision applies. Note that Sec. 250-148 requires that where multiple equipment grounding conductors enter a box, all must be spliced within the box or to the box so collective continuity is achieved.

SEC. 250-146–CONNECTING RECEPTACLE GROUNDING TERMINAL TO BOX

An equipment bonding jumper must be used to connect the grounding terminal of a grounding type receptacle to a grounded box. There are, however, several exceptions to this rule:

(a) **Surface-mounted boxes.** Where the box is mounted on or at the surface, direct metal-to-metal contact between the device yoke and the box is permitted to ground the receptacle to the box. There must be direct metal-to-metal contact in order to take advantage of this provi-

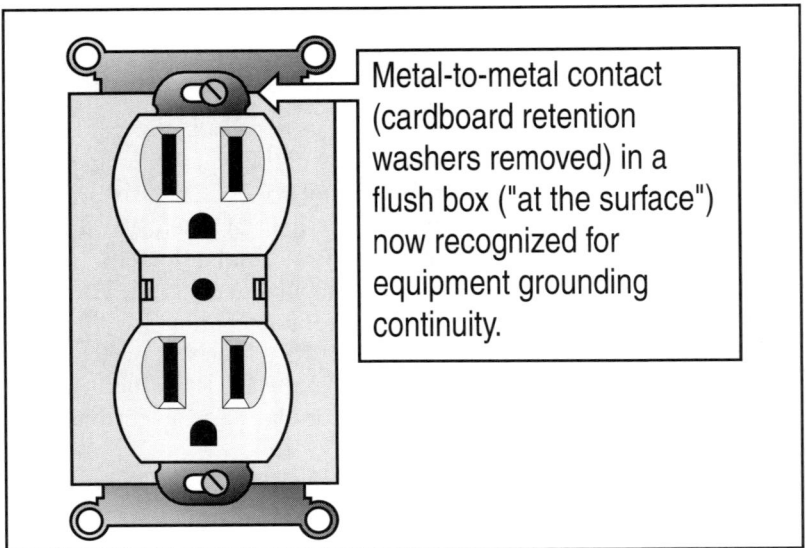

Metal-to-metal contact (cardboard retention washers removed) in a flush box ("at the surface") now recognized for equipment grounding continuity.

Fig. 7.7 This rule depends on direct metal-to-metal contact for proper operation, whether on a surface or flush at the surface.

sion, so be sure to remove any fiber retention washers from the mounting screws. See **Fig. 7.7.** If the front edge of the box is recessed, even a little, then it isn't "at" the surface, and you'll need a bonding jumper or a self-grounding yoke as covered in the next subsection.

This provision does not apply to cover-mounted receptacles unless the box and cover are listed as providing satisfactory ground continuity between the box and the receptacle. Although there are some hazardous-location and similar assemblies listed for this purpose, there are (as this is written) no conventional raised covers so listed. Any receptacle installed in a raised cover, therefore, needs a bonding jumper to be correct.

(b) Flush type boxes. Contact devices or yokes listed for the purpose, in conjunction with the supporting screws, are allowed to establish the grounding circuit between the device yoke and the flush-type box. These devices are often referred to as "self-grounding" devices.

(c) Floor boxes. Subsection (c) allows the use of floor boxes listed as providing satisfactory ground continuity between the box and the device.

(d) **Isolated ground receptacles.** Where required for the reduction of electrical noise on the grounding circuit, Section (d) allows the use of a receptacle in which the grounding terminal is purposely insulated from the receptacle mounting means. The grounding terminal of these isolated-ground (IG) receptacles must have an *insulated* equipment grounding conductor that is run with the circuit conductors back to, and terminated at, the equipment grounding conductor terminal of upstream distribution equipment. Generally, how far upstream the isolation is maintained is a design decision.

Under *no circumstances* is this permission either to run the grounding conductor *anywhere except* with the circuit conductors, or to terminate the grounding conductor *anywhere except* on a terminal that is part of a low-impedance ground return path for the same circuit, or for its parent feeder.

The insulated equipment grounding conductor from the IG receptacle is allowed to pass through one or more panelboards on its way to the service or derived system without connecting to the grounding terminals within these panelboards, as permitted in 384-20. However, the isolation must not be carried further upstream than a building disconnecting means **(see Fig. 7.8).**

Note that the raceway system and outlet box in which the IG receptacle is mounted must be grounded. Obviously, the insulated equipment grounding conductor does not accomplish this task, so another means of grounding the raceway and box must be found. You could use a metal raceway that independently qualifies as an equipment grounding conductor (See Sec. 250-118, covered in Chapter 8 of this book). In addition, some Type MC cable is available with two equipment grounding conductors in it, a green one for the box, and a green conductor with yellow stripes for use with the isolated grounding terminal.

Note that per Sec. 410-56(c), these receptacles must be identified with an orange triangle on the face. Normally these receptacles have to go in a metal box if you're using metal faceplates. However, Sec. 410-56(c) Ex. does allow isolated ground (IG) receptacles in a nonmetallic box with a metal faceplate, provided the box contains a feature that allows the yoke to be effectively grounded.

The exception recognizes designs under development, including some in other countries, that have at least one of the 6-32 mounting screws land in a metallic plate or other conductive area that leads to a grounding terminal. For IG receptacles, the only grounding contact for the yoke and therefore the faceplate is through those mounting screws, since the grounding terminal of the receptacle is intentionally insulated from the yoke. The grounding connection must be there in order to safely use a metal faceplate.

Faceplate grounding. Both Art. 410 and 380 have important ground-

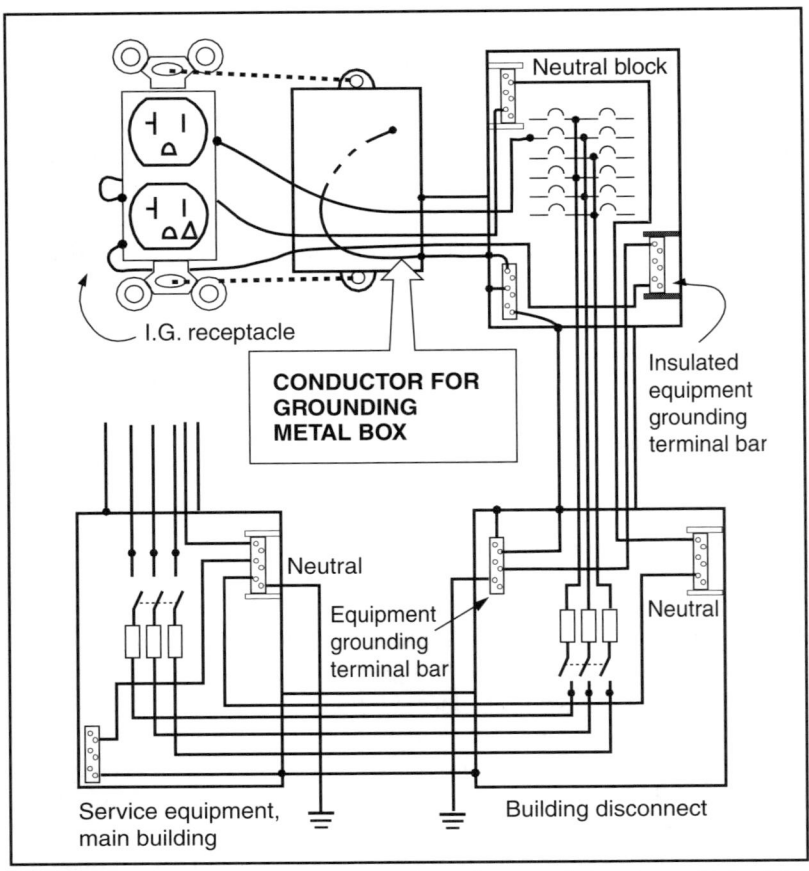

Fig. 7.8 Metal box containing IG receptacle can be grounded as shown. Maintaining the isolation upstream of a building disconnect is no longer permitted.

ing rules that address metal faceplates. Sec. 410-56(d) simply requires all metal receptacle faceplates to be grounded, and that will be easy or difficult depending on whether or not the yoke is grounded. This comes up on isolated ground receptacles in plastic boxes, and also on nongrounding or GFCI receptacle replacements for old receptacles with no equipment grounding conductor available, as covered in Sec. 210-7(d)(3).

Sec. 380-9 requires all snap switch yokes to be grounded, including dimmer switches, even if there is a nonmetallic faceplate. The only exception, generally recognizing a nonmetallic faceplate, is a very limited one for applications with no equipment ground available. The requirement reflects a simple reality, namely, that unqualified end users change out plastic plates for metal ones without permits or inspections, and the NEC has to anticipate that outcome.

Don't confuse the grounding requirements for switch yokes with those for receptacles. All you need is the mounting screws in the case of metal boxes; you don't need "self-grounding" switch yokes, etc. The self-grounding provisions of Section 250-146(b) only apply to receptacles. Receptacles, while in use, are a link in the branch-circuit equipment grounding return path and the yoke grounding connection is critical if it is being relied upon. This is not true for switches.

SEC. 250-130(C)–NONGROUNDING RECEPTACLE REPLACEMENTS OR BRANCH CIRCUIT EXTENSIONS

There is a special rule to accommodate the case where a nongrounding receptacle energized through a 2-wire branch circuit that does not include a grounding conductor is replaced with a grounding-type receptacle. The grounding conductor from the receptacle can be connected to any point on the grounding electrode system, including the nearest effectively grounded building column, or other means to which the equipment grounding conductor has been bonded at the service equipment.

At one time, as shown in **Fig. 7.9**, this also meant a nearby metallic water pipe. Now, with the limitation in the NEC of the use of a metallic water pipe as a conductor to that portion within 5 ft. of its entrance into the building, this is no longer applicable in residential occupancies. In a

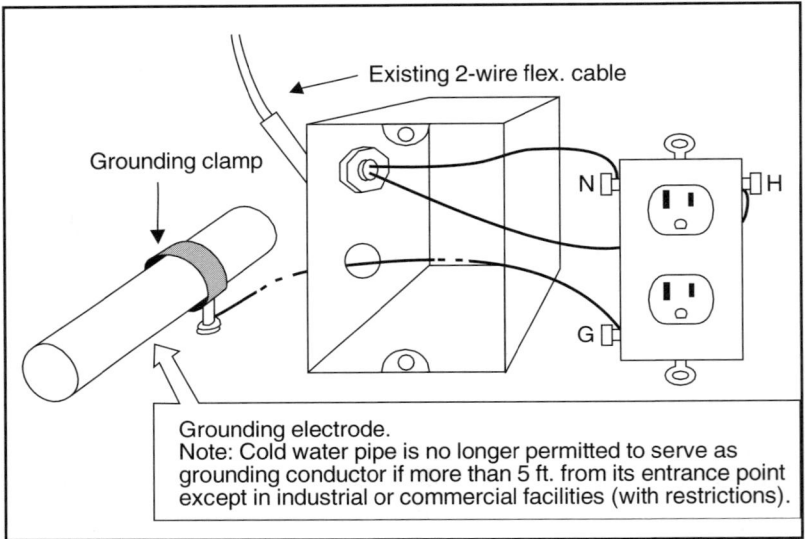

Existing 2-wire flex. cable

Grounding clamp

N H

G

Grounding electrode.
Note: Cold water pipe is no longer permitted to serve as grounding conductor if more than 5 ft. from its entrance point except in industrial or commercial facilities (with restrictions).

Fig. 7.9 Where no equipment grounding conductor is available, an exception allows the replacement of a nongrounding-type with a grounding-type receptacle by connecting the grounding screw to the nearest grounding electrode. A water-pipe connection (unless within 5 ft. of its entrance to the structure) is no longer permitted in residential occupancies.

metal-frame building, this is not a problem. In a wood-frame building, however, it virtually eliminates this exception unless the grounding connection is made within 5 ft of the entrance of the water pipe into the structure, or you qualify under the exception for remote water pipe grounding connections.

Even if you can get the grounding wire down to the water pipe entrance, you probably wouldn't use this provision anyway. The reason is simple: fish wires have two ends. If you can get the fish down to the basement, then you can get a new circuit using a modern wiring method back up. That generally makes much more sense.

An FPN refers to sec. 210-7(d) for information about how the replacement of GFCI receptacles is to be accomplished.

ARTICLE 250, PART F — EQUIPMENT GROUNDING AND EQUIPMENT GROUNDING CONDUCTORS

8

This chapter covers the basic "nuts and bolts" requirements for equipment grounding conductors — what they can be made of, how they need to be installed, and how we assure continuity. We covered the first four sections of this part in Chapter 6, and we won't repeat that coverage now. In addition, we are covering Sec. 250-148, the last section in Part G, in this chapter. It closely relates to Sec. 250-124 in this part. Sizing of these items is not covered in this chapter; all these requirements and the tables needed for determining the proper size of conductors are grouped together in Chapter 11 of this book. Sample calculations for typical installations are also contained there.

SEC. 250-118 – MATERIAL

The material from which the equipment grounding conductors can be made are spelled out in the following subsections.

(a) **Equipment grounding conductor types**. Equipment can be grounded by any of the grounding conductors listed in Sec. 250-118. These include:

- Copper or other corrosion-resistant conductor;
- Rigid metal conduit;
- IMC;
- EMT;
- Flexible metal conduit where both the conduit and fittings are listed for grounding
- Listed flexible metal conduit that is not listed for grounding but meeting certain conditions:

(1) The conduit fittings are listed for grounding;

(2) The overcurrent device(s) ahead of the circuit conductors in the

125

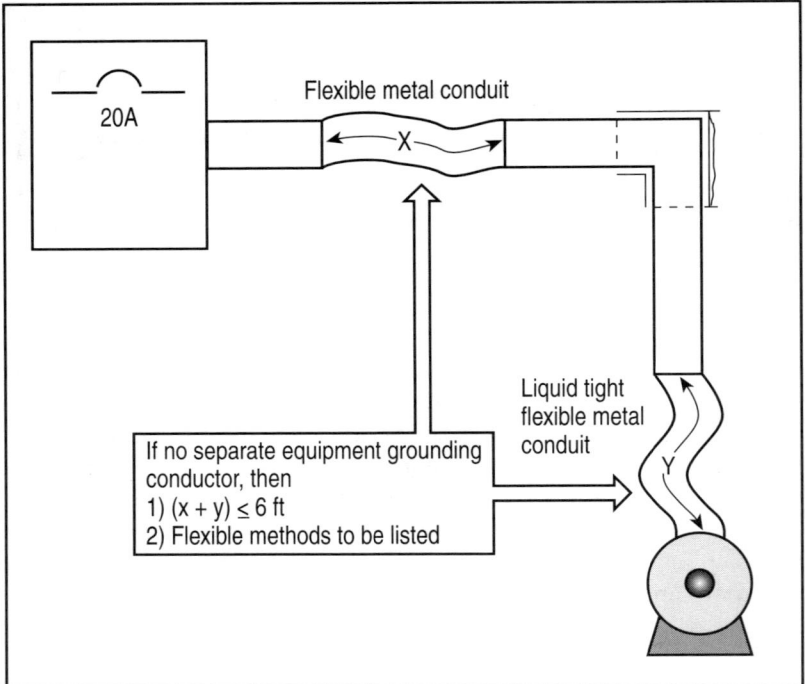

Fig. 8.1 The entire length of a flexible equipment grounding return path must be considered.

conduit is rated at 20 amperes or less;

(3) The combined length of flexible metal conduit plus flexible metallic tubing plus liquidtight flexible metal conduit in the same ground return path does not exceed 6 ft. (1.83 m)(see **Fig. 8.1**); and

(4) The conduit is not installed for flexibility.

• Listed liquidtight flexible metal conduit that is not listed for grounding but meeting certain conditions:

(1) The fittings are listed for grounding;

(2) For 3/8 in. - ½ in. trade-size conduit, the circuit overcurrent device(s) ahead of conductors in the conduit is rated at 20 amperes or less;

(3) For ¾ in. - 1¼ in. conduit, the circuit overcurrent device is rated at 60 amperes or less;

(4) The combined length of flexible metal conduit plus flexible metallic tubing plus liquidtight flexible metal conduit in the same ground return path does not exceed 6 ft. (1.83 m); and

(5) The conduit is not installed for flexibility.

• Flexible metallic tubing where the tubing is terminated in fittings listed for grounding and meeting certain conditions:

(1) The circuit overcurrent device(s) ahead of conductors in the tubing is rated at 20 amperes or less; and

(2) The combined length of flexible metal conduit plus flexible metallic tubing plus liquidtight flexible metal conduit in the same ground return path does not exceed 6 ft. (1.83 m).

• The armor of Type AC cable;

• The copper sheath of Type MI cable;

• The metallic sheath or combined metallic sheath and grounding conductors of Type MC cable;

• Cable trays, as permitted by Sec. 318-3(c) and 318-7;

• Cablebus, as permitted by Sec. 365-2(a); and

• Other continuous metal raceways listed for grounding.

(b) With circuit conductors. Equipment can also be grounded by a grounding conductor contained within the same raceway, cable, or cord, or otherwise run with the circuit conductors.

The referenced section of Art. 318 requires the cable tray used as an equipment grounding conductor to have a certain minimum cross-sectional area of metal. Table 318-7(b)(2) is used for determining this. The table and sample calculations are included in Chapter 11 of this book.

Cable trays used for this purpose are limited to commercial and industrial establishments that have continuous maintenance and supervision to assure that only qualified persons will service the installed cable tray system. In addition, metallic cable trays themselves must be grounded per the rules spelled out in Article 250.

Steel or aluminum cable tray systems used for this purpose must meet all of the following requirements;

• The cable tray sections and fittings must be identified as suitable for grounding purposes.

• As mentioned previously, the tray must have a minimum cross-sectional area;

• All cable tray sections and fittings must be legibly and durably marked to show the cross-sectional area of metal in channel cable trays (cable trays of one-piece construction), and the cross-sectional area of both side rails for ladder or trough cable trays.

• Cable tray sections, fittings, and connected raceways must be bonded in accordance with Sec. 250-96 using bolted mechanical connectors or bonding jumpers sized in accordance with 250-102. Both of these are covered in Chapter 3 of this book.

Sec. 365-2(a) states that cablebus framework, where adequately bonded, is permitted to be used as the feeder grounding conductor for branch circuits and feeders. Further, Sec. 365-9 says that the sections of cablebus must be electrically bonded either by inherent design of the mechanical joints or by applying bonding means as required by Sec. 250-96. The cablebus installation, in turn, must be grounded as required by Secs. 250-80 and 250-86.

Note that Sec. 250-86 Ex. 2 is specifically excluded from applicability, and therefore the allowance for short raceways sections to remain ungrounded if used as a support for cable assemblies cannot be applied. The cablebus support frames must be grounded.

Sec. 250-119 – Identification of equipment grounding conductors. Equipment grounding conductors can be bare, covered, or insulated. If covered or insulated, the material must have a continuous outer finish that is either green or green with one or more yellow stripes. There are several modifications to this color-coding requirement.

(a) Conductors larger than No. 6. These conductors can be identified by marking the terminations and at every intervening point where the conductor is accessible, even conduit bodies. You either strip bare the exposed length, or you apply green tape or adhesive labels, or you color the conductor itself.

(b) Multiconductor cable. If, and only if, there is qualified maintenance and supervision over the installation, you can permanently re-identify one of the contained conductors, even if smaller than No. 6, in the same way as is generally allowed for No. 6 and larger conductors, and at the same points in the wiring system.

(c) Flexible cord. An uninsulated conductor is permitted by this rule, along with the traditional green or green-with-yellow-stripes insulation.

Note that Sec. 400-23 effectively requires insulated (or at least braid-covered) conductors, however.

SEC 250-120 – INSTALLATION

Equipment grounding conductors must be installed as specified in the following subsections.

(a) Where it consists of a raceway, cable tray, cable armor, or cable sheath, or where it is a wire within a raceway or cable, the equipment grounding conductor must be installed in accordance with the applicable provisions in the NEC. In the installation, fittings for joints and terminations that are approved for use with the type of raceway or cable must be used, and all connections, joints, and fittings must be made tight using suitable tools.

(b) Aluminum and copper-clad aluminum conductors must meet the same restrictions against proximity to soil and masonry as grounding electrode conductors made of the same material, as covered in Sec. 250-64 (Chapter 2 in this book).

(c) Equipment grounding conductors smaller than No. 6 must be protected from physical damage by a raceway or cable armor. Sizes smaller than No. 6, however, are not required to be enclosed in a raceway or armor where run in hollow spaces of a wall or partition, or where otherwise installed so that they are not subject to physical damage. One place this issue comes up is in the routing of grounding conductors from replacement receptacles on systems without equipment grounding conductors, as covered in Sec. 250-130(c).

SEC 250-124 – EQUIPMENT GROUNDING CONDUCTOR CONTINUITY

The following subsections contain rules for maintaining continuity of the equipment grounding conductor at plug-in devices, switches, and other similar devices.

(a) Separable connections such as those provided in draw-out equipment or attachment plugs and mating connectors or receptacles must provide for first-make, last-break of the equipment grounding conductor.

An exception to this rule is made for interlocked equipment, plugs,

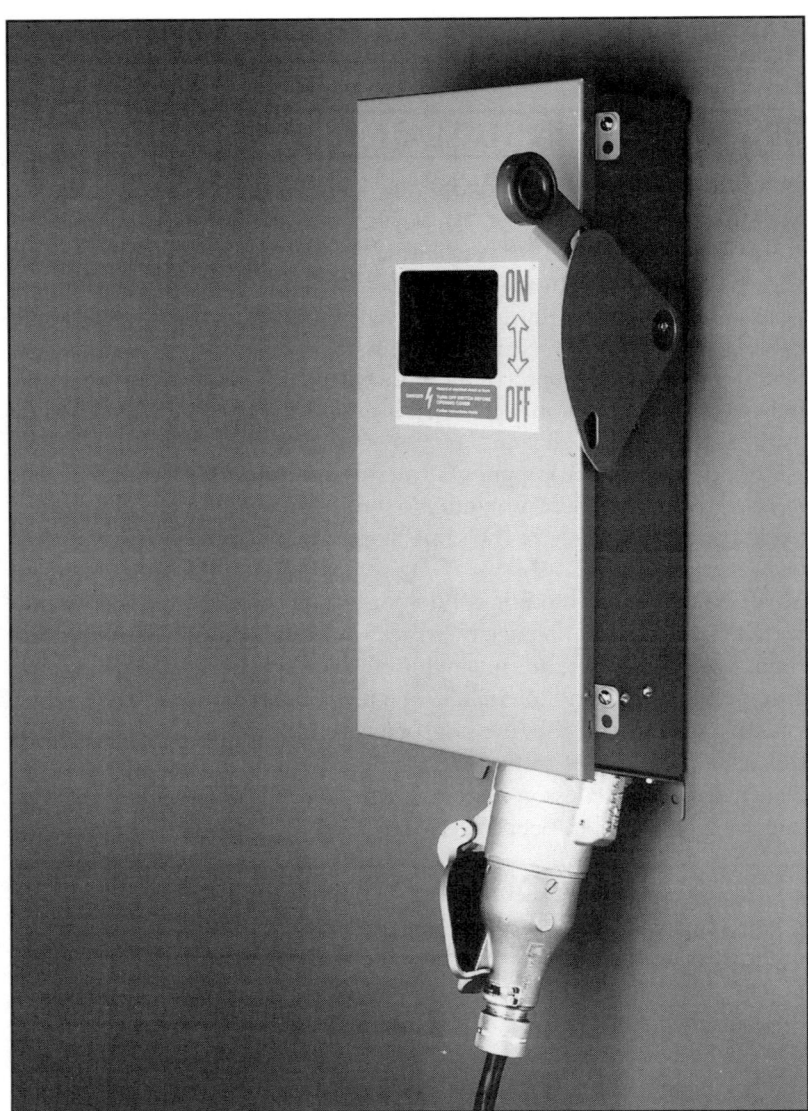

Interlocking mechanisms that prevent withdrawal of the plug while the switch is in the ON position are considered as assuring that continuity of the equipment grounding conductor is maintained while the equipment being fed is in use.

receptacles, and connectors that preclude energization without grounding continuity.

(b) Switches. No automatic cutout or switch is allowed to be placed in the equipment grounding conductor of a premises wiring system.

An exception is made where the opening of the cutout or switch disconnects all sources of energy.

SEC. 250-148–CONTINUITY AND ATTACHMENT OF BRANCH-CIRCUIT GROUNDING CONDUCTORS TO BOXES

Where more than one equipment grounding conductor enters a box, all such conductors are required to be spliced or joined within the box or to the box with devices suitable for the use. Connections depending solely on solder are not to be used for the purpose.

Listed bolted split-type connectors provide an easy and reliable way to make connections.

An exception says that the equipment grounding conductor permitted in Sec. 250-146(d) is not required to be connected to the other equipment grounding conductors or to the box where an "isolated-ground receptacle" is installed. In such instances, the receptacle grounding terminal is isolated from the receptacle mounting and an insulated equipment grounding conductor is run with the circuit conductors back through one or more intervening panelboards.

As shown in **Fig. 8.2**, the arrangement of grounding connections must be such that the disconnection or the removal of a receptacle, fixture, or other device fed from the box will not interfere with or interrupt the grounding continuity.

Splices are to be made in accordance with Sec. 110-14(b). The rule there is entitled "splices" and says:

> *Conductors shall be spliced or joined with splicing devices identified for the use or by brazing, welding, or soldering with a fusible metal or alloy. Soldered splices shall first be so spliced or joined as to be mechanically and electrically secure without solder and then soldered.*

Sec 110-14(b) then continues to say that the splice must be insulated. In the situation covered by Sec. 250-148 for grounding conductors, insulation is not required.

Two subsections to Sec. 250-148 contain rules that apply to boxes depending upon their materials of construction.

(a) Metal boxes. A connection must be made between the one or more equipment grounding conductors and a metal box by means of a grounding screw that is to be used for no other purpose (see **Fig. 8.2**, referenced earlier), or a listed grounding device, such as a "G-clip."

(b) Nonmetallic boxes. One or more equipment grounding conductors brought into nonmetallic outlet box must be so arranged that a connection can be made to any fitting or device in that box requiring grounding.

In general, nonmetallic boxes must be used with non-metallic wiring methods, as covered in Sec. 370-3, but there are two exceptions:

Exception No. 1 permits metallic wiring methods with "internal" bonding. This can be accomplished with grounding hubs and/or bushings properly connected to the entering wiring method(s), and the use of appropriately sized bonding jumpers between them within the box.

Exception No. 2 permits metallic wiring methods to be connected to nonmetallic boxes of any size with "integral" bonding – usually a metal mesh or framework molded throughout the walls of the box. In addition, an integral method must be provided by the manufacturer to allow the attachment of bonding jumpers should equipment mounted within the box require grounding.

Sec. 250-126 – Identification of wiring device terminals

The terminals for the connection of the equipment grounding conductor to a device must be identified by:

(1) A green-colored, not readily removable terminal screw with a hexagonal head;

(2) A green-colored, hexagonal, not readily removable terminal nut; or

(3) A green-colored pressure wire connector.

If the terminal for the grounding conductor is not visible, the conductor entrance hole must be marked with the word "green" or "ground" or

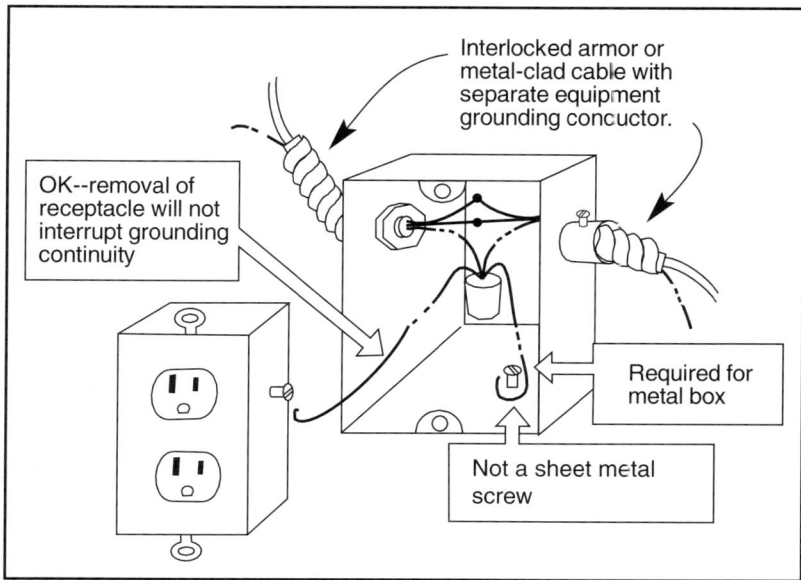

Interlocked armor or metal-clad cable with separate equipment grounding conductor.

OK--removal of receptacle will not interrupt grounding continuity

Required for metal box

Not a sheet metal screw

Fig. 8.2 One of the keys to maintaining the integrity of the grounding path is to make sure that removal of receptacles and the like from boxes does not interrupt the equipment grounding conductor.

the letters "G" or "GR" or the inverted tree symbol universally used to denote grounding (refer to **Fig. 1.1** in this book for an example). Otherwise the terminal must be identified by a distinctive green color. If the terminal is readily removable, then the area adjacent to the terminal needs to be similarly marked.

ARTICLE 250, PART J — INSTRUMENTS, METERS AND RELAYS

Metering and instrumentation play significant roles in electrical distribution and control systems. While generally covered by the rules in other parts of Article 250, this is specialized equipment that merits a more detailed explanation of how these rules are to be applied to cover them. Part J of Article 250 addresses this need.

The wording in this section is somewhat ambiguous. Window-type CT's

Instrument transformers are associated with ammeters, voltmeters, and other monitoring and instrumentation circuits mounted on control panels of centralized control rooms.

are not "connected to" the circuit in which they monitor current. Within the intent of the wording, however, the requirements that apply when a wound-primary CT is used is also applicable to the window-type CT.

Similarly, the word "switchboard" is used throughout this part. In the requirements stated in Part J, this term can be taken to mean a panel in or on which CT's, PT's, and instrumentation and metering equipment are mounted. This can be on a switchboard, in drawout switchgear compartments, in a box, or within a control panelboard.

SEC. 250-170–INSTRUMENT TRANSFORMER CIRCUITS

The secondary circuits of instrument current transformers (CT's) are required to be grounded where the primary windings are connected to circuits of 300V or more to ground (see **Fig 9.1**). A similar provision applies to potential transformers (PT's). When these units are mounted on switchboards, they must be grounded regardless of the voltage level of the primary.

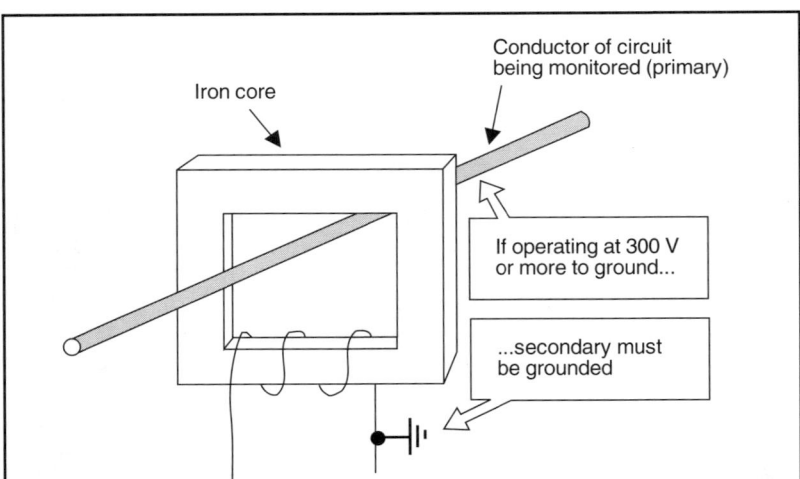

Fig. 9.1. A window-type CT consists of an iron core through which the conductor of the circuit being monitored runs. The secondary winding wound around the core normally carries approximately 5A when the current in the primary is at its maximum

An exception to this rule eliminates the need for grounding the secondary circuits when:
- The primary windings are connected to circuits of less than 1000V; and
- No live parts or wiring is exposed or accessible to other than qualified persons.

This is one of the cases where the terms used in the text obscure the actual requirements. CT's and PT's are often mounted on or within control panelboards (switchboards) in central control rooms or in manufacturing areas. These locations are often accessed by process operators, non-electrician maintenance personnel, and others. Thus, the terminals of CT's and PT's mounted on them are more likely to be contacted and, thus, must have their secondaries grounded.

On the other hand, where location and training make such items accessible only to qualified persons, such as when they are located in a switchgear metering compartment, the secondaries are not required to be grounded as long as they are connected to circuits operating at less

Trained electricians monitor and service instruments located on switch-gear and, thus, grounding the secondary of a CT or PT is not mandatory, but sti l desirable.

than 1000V. This exception, however, does not prohibit such grounding. It is generally prudent to ground the secondary circuits involved.

SEC. 250-172 – INSTRUMENT TRANSFORMER CASES

Cases or frames of instrument transformers must be grounded where accessible to other than qualified persons. This requirement is in addition to the previous one for grounding of the secondary circuits.

An exception is made to the grounding requirement for cases or frames of CT's (not PT's) having primaries that are not over 150V to ground and the CT's are used exclusively to supply current to meters (**Fig 9.2**).

SEC. 250-174 – CASES OF INSTRUMENTS, METERS AND RELAYS OPERATING AT LESS THAN 1000V

Besides the instrument transformer, Part L also requires grounding the

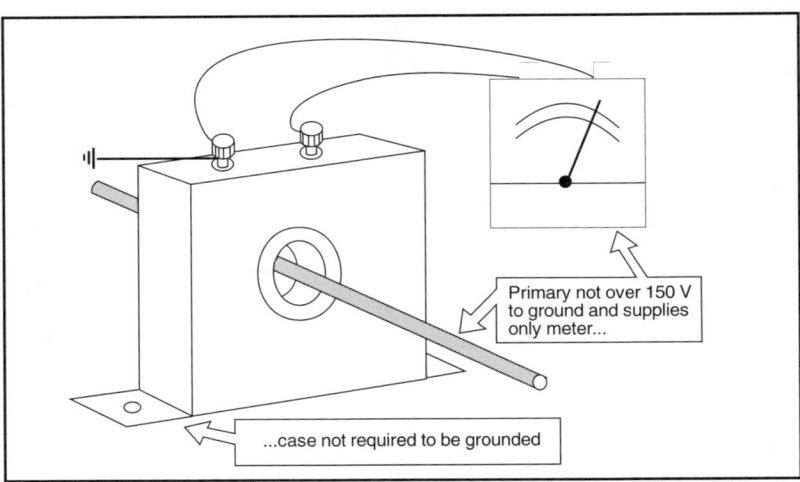

Primary not over 150 V to ground and supplies only meter...

...case not required to be grounded

Fig. 9.2. An exception to the rule for grounding the case of instrument transformers is made for a CT where the primary is not over 150V to ground, and the secondary only supplies a meter.

case of instruments, meters, relays, and similar devices. Grounding is required for those items having windings or working parts operating at less than 1000V. The following subsections give the details.

(a) Not on switchboards. Instruments, meters, and relays not located on switchboards, operating with windings or working parts at 300V or more to ground, and accessible to other than qualified persons, must have the cases and other exposed metal parts grounded.

(b) On dead-front switchboards. Instruments, meters, and relays (whether operated from CT's, PT's, or connected directly in the circuit) on switchboards having no live parts on the front of the panels must have their cases grounded if the cases are made of metal.

This description fits most installations of this type of equipment, either mounted as individual units or grouped together on control consoles and similar locations (see **Fig. 9.3**).

(c) On live-front switchboards. Instruments, meters, and relays (whether operated from CT's, PT's, or connected directly in the circuit) on switchboards having exposed live parts on the front of panels are prohibited from having their cases grounded. A mat of insulating rubber

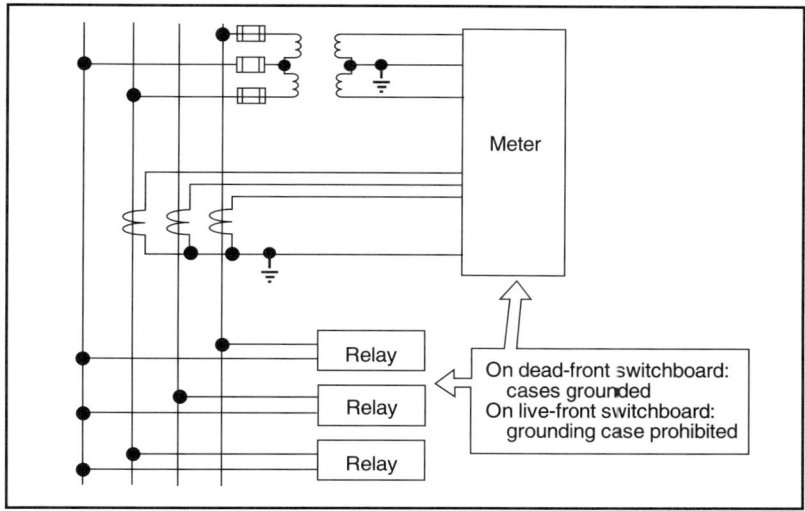

Fig. 9.3 Grounding of the case of instruments, relays, etc. (if of metal) is required if they are mounted on deadfront switchboards

or other suitable floor insulation is to be provided for the operator where the voltage to ground exceeds 150V.

This rule applies mainly to older installations where open knife switches or other live parts are mounted on the front of a panel made of insulating material. These panels often also contained ammeters, voltmeters, recording instruments, and similar equipment. If the cases of the instruments are grounded, someone who inadvertently simultaneously contacts a piece of live equipment and an instrument case would receive a shock. To prevent such an occurrence, the operator must be *insulated* from ground by an appropriate mat, and by *not* grounding the instrument cases.

SEC. 250-176–CASES OF INSTRUMENTS, METERS AND RELAYS OPERATING AT 1kV AND OVER

Where instruments, meters, and relays have current-carrying parts of 1kV and over to ground, they must be isolated by elevation or protected by suitable barriers, grounded metal, or insulating covers or guards. Their cases must not be grounded.

An exception allows cases of electrostatic ground detectors to be grounded where the internal ground segments of the instrument are connected to the instrument case and grounded and the ground detector is isolated by elevation.

SEC. 250-178–INSTRUMENT GROUNDING CONDUCTOR

The grounding conductor for secondary circuits of instrument transformers and for instrument cases must not be smaller than No. 12 copper or No. 10 aluminum.

Cases of instrument transformers, instruments, meters, and relays that are mounted directly on grounded metal surfaces of enclosures or grounded metal switchboard panels are considered to be grounded; and no additional grounding conductor is required.

ARTICLE 250, PART K — GROUNDING OF SYSTEMS AND CIRCUITS OF 1kV OR OVER (HIGH VOLTAGE)

The title of this article is something of a misnomer since most of the circuits covered by the Code operate at medium voltage (1kV to 69 kV). High-voltage systems (over 69kV) are most often electric utility-operated generation and transmission and distribution installations, which are not covered by the NEC. Customer-owned high-voltage distribution systems, however, are covered by Article 490 and those parts of other NEC articles that cover over 600V installations.

Rules for grounding given in this chapter apply from the point where

Monitoring and instrumentation devices on this lineup of MV metalclad switchgear are in contact with the grounded panels and are, thus, not required to be individually grounded by an equipment grounding conductor.

the electrical utility's service conductors terminate. Many types of installations fall under the jurisdiction of this set of rules. Mobile equipment such as sled-mounted unit substations, large gantry cranes, and shore-fed dredges are affected. Fixed installations such as offshore drilling/offloading platforms and medium-voltage (or high-voltage) distribution systems within large manufacturing complexes, and high rise buildings are also covered.

SEC. 250-180–GENERAL

All of the rules applying to systems and circuits operating at lower voltages also apply to systems and circuits operating at voltages equaling and exceeding 1kV. The set of rules presented in this part only supplement or modify those general rules.

This parallels what is said in Sec. 250-20(c) which permits (but does not require) AC systems of 1kV and over to be grounded. This section states that when these systems are grounded, they must comply with other applicable provisions of Article 250.

Medium-voltage systems that distribute power within large manufacturing facilities are covered by Part K of the NEC.

SEC. 250-182 – DERIVED NEUTRAL SYSTEMS

Code rules on grounding of medium-voltage circuits allow a system neutral (when one is needed) to be derived through a grounding transformer.

For instance, large motors of mobile equipment often are connected to a system operating at a medium voltage, such as 2400V, 4160V, and 13.8 kV. The power distribution system at these higher voltages are sometimes derived from delta-connected transformers. Thus, no system neutral is available. As is required in Sec. 250-154, however, this type of equipment must be fed from an impedance-grounded power distribution system. To comply with this rule, it is necessary to use a grounding transformer to derive a grounded (neutral) conductor.

Rules that apply to grounding transformers are found in the NEC Sec. 450-5. The subject is discussed in a companion book to this one, entitled Understanding NE Code Rules on Transformers.

SEC. 250-184 – SOLIDLY GROUNDED NEUTRAL SYSTEMS

Medium- or high-voltage, solidly-grounded systems distributing power over a wide area are often in the form of 4-wire, multiple-ground, common-neutral systems such as the one shown in **Fig. 10.1**. The rules covering these and similar types of installations are spelled out in the following subsections.

(a) Neutral conductor. When a 1kV or higher circuit is solidly grounded, the insulation level of the neutral conductor is required to be 600V minimum. Note, however, that the insulation level is not required to be higher - although it can be higher.

There are two exceptions to the requirement that the neutral be insulated. Bare copper conductors are permitted to be used for the neutral of:

• Service entrances and direct buried portions of feeders; and;

• Overhead portions installed outdoors (although it need not be copper).

An FPN refers to Sec. 225-4 for requirements where the neutral conductor is within 10 ft. (3.05 m) of any building or other structure. There it is required that such conductors be insulated or covered. An excep-

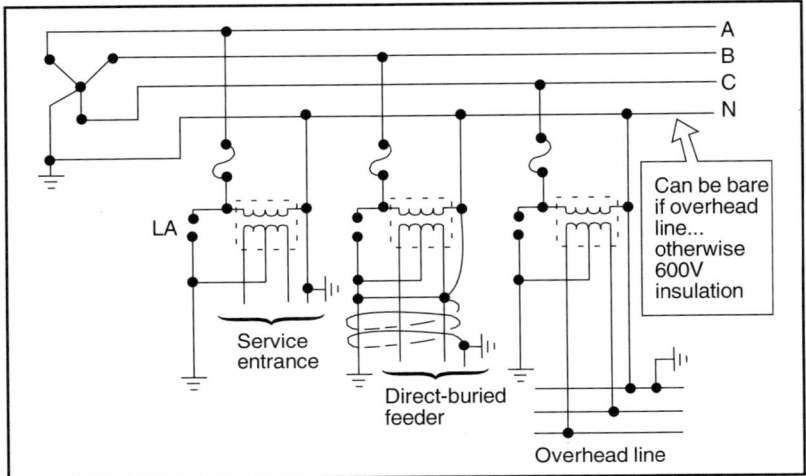

A
B
C
N

Can be bare if overhead line... otherwise 600V insulation

LA

Service entrance

Direct-buried feeder

Overhead line

Fig. 10.1 A typical MV 4-wire, multiple ground, common-neutral distribution system.

tion, however, states that grounded circuit conductors are permitted to be bare when they are permitted to be so elsewhere in the Code. The permission granted by the exceptions to Sec. 250-184 (a), thus, is still valid even when the conductor is within 10 ft. of a building or structure.

SEC. 250-184(b)-MULTIPLE GROUNDING

The neutral of a solidly-grounded neutral system can be grounded at more than one point at a service. This rule coordinates with that of Sec. 250-24, where it is required that if the transformer supplying the service is located outdoors, at least one additional grounding connection must be made from the grounded service conductor to a grounding electrode, either at the transformer or outside of the building. For more information on this subject, see Chapter 5 in this book.

The neutral of a solidly-grounded neutral system may also be grounded in more than one point at direct-buried portions of feeders employing a bare copper neutral, and at overhead portions of a circuit installed outdoors. In each of these cases, supplementary ties to the grounding electrodes can be made along the length of the run.

Sec. 250-184(c) The neutral grounding conductor is permitted to be of bare copper if isolated from phase conductors and protected from physical damage. Previous sections of Article 250 refer to the terms "neutral conductor" and "grounding electrode conductor", but no mention is made of "neutral grounding conductors". In a multiple grounded system, as shown in **Fig. 10.2**, it describes the neutral tap from an overhead (or direct buried) line that extends to a grounding electrode and is used both as a neutral and to ground various components.

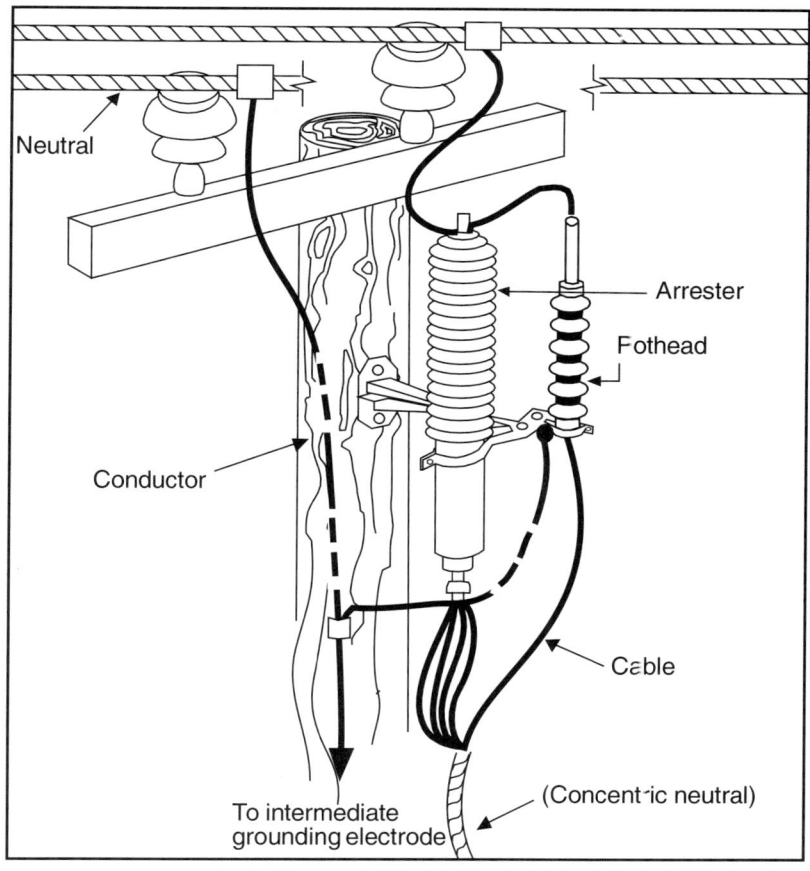

Fig. 10.2 The neutral grounding conductor of a solidly-grounded MV or HV system is permitted to be bare if protected from damage.

SEC. 250-186–IMPEDANCE GROUNDED NEUTRAL SYSTEMS

These systems, shown in **Fig. 10.3**, are required by Sec. 250- 188 for portable or mobile equipment. When medium voltage impedance-grounded neutral systems are used, they must comply with provisions of the following subsections. These systems are usually classified as either high- or low-impedance, as explained in Chapter 5.

More often than not, however, these are low impedance systems, arranged with fault relaying to disconnect power quickly instead of the monitoring and continuity of power contemplated in high-impedance systems covered for 480Y/277V and 600Y/347V systems in Sec. 250-36. In these systems the impedance serves to lower the amount of fault current, mitigating the damage that would otherwise be caused by the amount of energy available in these systems.

(a) **Location.** The grounding impedance must be inserted in the grounding conductor between the grounding electrode of the supply sys-

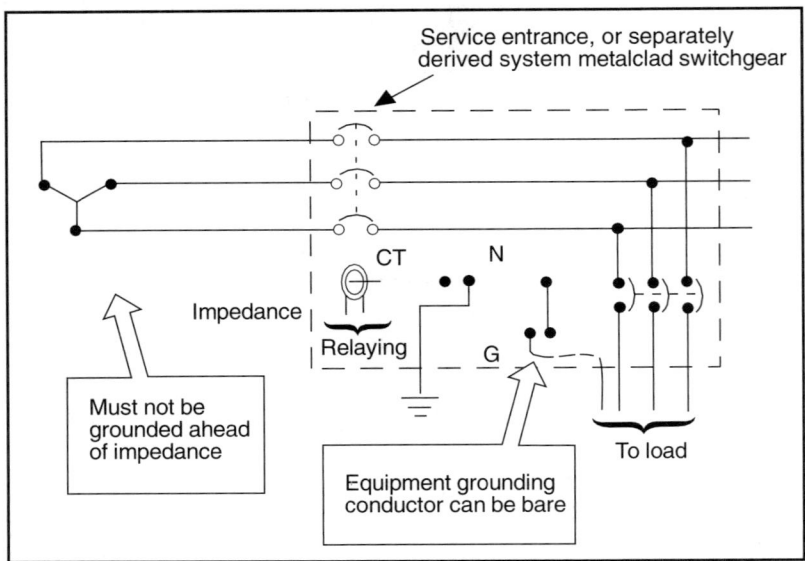

Fig. 10.3 An impedance-grounded system grounding arrangement.

tem and the neutral point of the supply transformer or generator.

(b) Identified and insulated. Where the neutral conductor of an impedance-grounded neutral system is used, it must be identified. Secs. 200-6 and 200-7 reserve the color white or natural gray for the identification of a grounded conductor.

In addition to identification, the neutral conductor must be fully insulated to the same voltage level as the phase conductors.

(c) System neutral connection. The system neutral is not permitted to be connected to ground except through the neutral grounding impedance.

(d) Equipment grounding conductors are permitted to be bare. They must be connected to the ground bus and grounding electrode conductor at the service-entrance equipment or at the origin of a separately derived system and extended to the system ground. With modern industrial distribution systems, these systems are frequently premises wiring, not originating at a service.

SEC. 250-188–GROUNDING OF SYSTEMS SUPPLYING PORTABLE OR MOBILE EQUIPMENT

Systems supplying portable or mobile medium-voltage equipment, other than substations installed on a temporary basis, must comply with the rules outlined in the following subsections.

Note that Sec. 90-2(b)(2) exempts "self-propelled mobile surface mining machinery and its attendant electrical trailing cable" from NEC coverage. The non-self-propelled equipment shown in Fig. 10.4, however, *must* comply with these NEC rules.

(a) Portable or mobile equipment must be supplied from a system having its neutral grounded through an impedance. Where a delta-connected medium-voltage system is used to supply portable or mobile equipment, a system neutral *must* be derived.

The rules that were outlined in the discussion of Sec. 250-186 are, thus, applicable to the portable and mobile equipment covered by this section.

(b) Exposed noncurrent-carrying metal parts of portable or mobile equipment must be connected by an equipment grounding conductor to

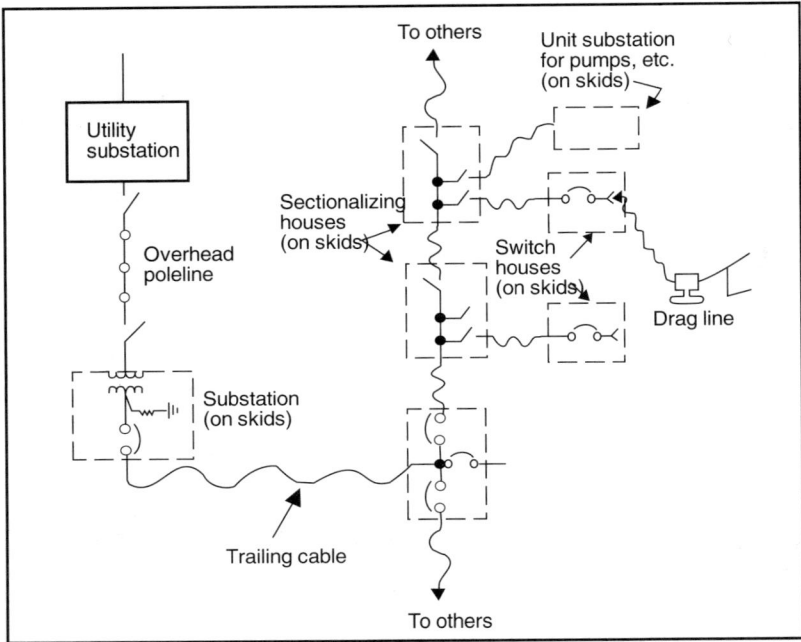

Fig. 10.4 Typical above-ground mining operation with MV drag line.

the point at which the system neutral impedance is grounded.

Electricity to equipment of this type is supplied via extra-flexible mining cable (trailing cable). As shown in **Fig. 10.5**, included in this cable is an equipment grounding conductor(s). All noncurrent-carrying metal parts onboard the drag line or other equipment, as well as the frame of the unit itself, are grounded by this conductor. At the main substation feeding this portable or mobile equipment, the equipment grounding conductor is usually connected to the ground bus of switchgear. Bonding that connects this bus to the neutral bus and the grounding electrode conductor completes the tie to the point at which the system neutral impedance is grounded. Remember, according to Sec. 250-186, the system neutral can only be grounded downstream of the grounding impedance.

(c) Ground fault current. The voltage developed between the portable or mobile equipment frame and ground by the flow of maximum ground-fault current must not exceed 100V. This requirement is usually met by sizing the neutral grounding resistor to limit the fault current to a certain level (say 25A), and limiting the resistance of the ground wire (say to 4 ohms).

(d) Ground-fault detection and relaying must be provided to automatically de-energize any medium- or high-voltage system component that has developed a ground fault.

The continuity of the equipment grounding conductor must be continuously monitored so as to de-energize automatically the feeder to the portable or mobile equipment upon loss of continuity of the equipment grounding conductor.

Trailing cable contains a separate "ground check" conductor. This conductor completes a DC circuit through the equipment grounding conductor. If the circuit is interrupted, detector relaying will trip the protective device for the power circuit to the mobile equipment. Note that this cable is now recognized in Art. 400 as well.

(e) Isolation. The grounding electrode to which the portable or mo-

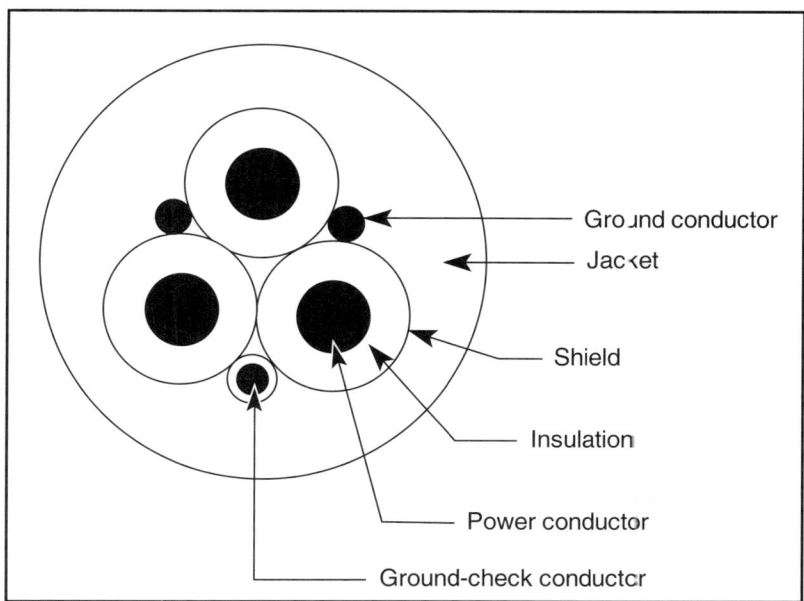

Fig. 10.5 Cross-section of a typical MV mining cable. Note the ground-check conductor.

bile equipment system neutral impedance is connected must be isolated from, and separated in the ground, by at least 20 ft. (6.1 m) from any other system or equipment grounding electrode, and there must be no direct connection between the grounding electrode and such items as buried pipe, fence, etc.

(f) Trailing cable and couplers for interconnection of portable or mobile equipment must been the requirements of Part C of Art. 400 for cables, and Sec. 490-55 for the couplers.

Article 400, Part C covers portable cables over 600V, nominal. The following topics in Part C contain information on the grounding of cables.

400-30. Scope. This part applies to multi-conductor portable cables used to connect mobile equipment and machinery.

400-31. Construction.

(a) Conductors. The conductors shall be No. 8 copper or larger and shall employ flexible stranding. The ground check conductor, however, can be as small as No. 10.

(b) Shields. Cables operated at over 2000 volts shall be shielded. Shielding shall be for the purpose of confining the voltage stresses to the insulation.

(c) Equipment Grounding Conductor(s). Equipment grounding conductor(s) shall be provided. The total area shall not be less than that of the size of the equipment grounding conductor required in Section 250-122.

400-32 Shielding. All shields shall be grounded.

400-34. Minimum Bending Radii. The minimum bending radii for portable cables during installation and handling in service shall be adequate to prevent damage to the cable.

400-35. Fittings. Connectors used to connect lengths of cable in a run shall be of a type that lock firmly together. Provisions shall be made to prevent opening or closing these connectors while energized. Suitable means shall be used to eliminate tension at connectors and terminations.

400-36. Splices and Termination. Portable cables shall not contain splices unless the splices are of the permanent molded, vulcanized types in accordance with Section 110-14(b). Terminations on portable cables rated over 600 volts, nominal, shall be accessible only to authorized and qualified personnel.

SEC. 490-55–POWER CABLE

Power Cable Connections to Mobile Machines. "A metallic enclosure shall be provided on the mobile machine for enclosing the terminals of the power cable. The enclosure shall include provisions for a solid connection for the ground wire(s) terminal to effectively ground the machine frame. Ungrounded conductors shall be attached to insulators or terminated in approved high-voltage cable couplers (which include ground wire connectors) of proper voltage and ampere rating. The method of cable termination used shall prevent any strain or pull on the cable from stressing the electrical connections. The enclosure shall have provisions for locking so only authorized and qualified persons may open it and shall be marked DANGER-HIGH VOLTAGE-KEEP OUT."

Fences and gates are among the many metallic objects that are required to be grounded.

SEC. 250-190-GROUNDING OF EQUIPMENT

All noncurrent-carrying metal parts of fixed, portable, and mobile equipment and associated fences, housings, enclosures, and supporting structures must be grounded, with the following exception.

Exception. Where isolated from ground and located so as to prevent any person who can make contact with ground from contacting such metal parts when the equipment is energized.

Grounding conductors not an integral part of a cable assembly must not be smaller than No. 6 copper or No. 4 aluminum. This rule includes bonding conductors from cable shields.

SIZING GROUNDING CONDUCTORS

Conductors used for grounding and bonding must be correctly sized so that they will provide an adequate low impedance path for fault current. There are many references scattered throughout Article 250 and the rest

Size of Largest Service-Entrance Conductor or Equivalent Area for Parallel Conductors[1]		Size of Grounding Electrode Conductor	
Copper	Aluminum or Copper-Clad Aluminum	Copper	Aluminum or Copper-Clad Aluminum[2]
2 or smaller	1/0 or smaller	8	6
1 or 1/0	2/0 or 3/0	6	4
2/0 or 3/0	4/0 or 250 kcmil	4	2
Over 3/0 thru 350 kcmil	Over 250 kcmil thru 500 kcmil	2	1/0
Over 350 kcmil thru 600 kcmil	Over 500 kcmil thru 900 kcmil	1/0	3/0
Over 600 kcmil thru 1100 kcmil	Over 900 kcmil thru 1750 kcmil	2/0	4/0
Over 1100 kcmil	Over 1750 kcmil	3/0	250 kcmil

Note 1: Where multiple sets of service-entrance conductors are used as permitted in Section 230-40, Exception No. 2 the equivalent size of the largest service-entrance conductor shall be determined by the largest sum of the areas of the corresponding conductors of each set.
Note 2: Where there are no service-entrance conductors, the grounding electrode conductor size shall be determined by the equivalent size of the largest service-entrance conductor required for the load to be served.
[1]This table also applies to the derived conductors of separately derived ac systems.
[2]See installation restrictions in Section 250-64.
FPN: See Section 250-24(b) for size of ac system conductor brough: to service equipment.

Fig. 11.1 Sizing grounding electrode conductors for ac systems. Based on Table 250-66 on the NEC.

of the NEC about sizing requirements for these conductors. Rather than treat with each individually in the sequence they appear in the Code, all of the pertinent information dealing with these calculations has been grouped into this chapter of the book.

There are two tables in Article 250 that are essential for the proper sizing of grounding and bonding conductors. They are:

Table 250-122.

Minimum Size Equipment Grounding Conductors for Grounding Raceway and Equipment

Rating or Settling of Automatic Overcurrent Device in Circuit Ahead of Equipment, Conduit, etc., Not Exceeding (Amperes)	Size	
	Copper Wire No.	Aluminum or Copper-Clad Aluminum Wire No.[1]
15	14	12
20	12	10
30	10	8
40	10	8
60	10	8
100	8	6
200	6	4
300	4	2
400	3	1
500	2	1/0
500	1	2/0
800	1/0	3/0
1000	2/0	4/0
1200	3/0	250 kcmil
1600	4/0	350 kcmil
2000	250 kcmil	400 kcmil
2500	400 kcmil	600 kcmil
3000	400 kcmil	600 kcmil
4000	500 kcmil	800 kcmil
5000	700 kcmil	1200 kcmil
6000	800 kcmil	1200 kcmil

(ROP 5-300, 301, 302)

Note: Where necessary to comply with Section 250-2(d), the equipment grounding conductor shall be sized larger than this table.

    [1]See installation restrictions in Section 250-120.

Fig. 11.2 Minimum size equipment grounding conductors for grounding raceways and equipment. Based on Table 250-122 in the NEC.

• Table 250-66 - Grounding Electrode Conductor for AC Systems. This is reproduced as **Fig. 11.1**; and

• Table 250-122 - Minimum Size Equipment Grounding Conductors for Grounding Raceway and Equipment. This is reproduced as **Fig. 11.2** in this book. This table goes with Sec. 250-122, that has many specialized sizing rules for equipment grounding conductors. Please refer to the end of this chapter for a complete discussion.

Another important table for sizing of equipment grounding conductors is found in Article 318. The table there, **Table 318-7(b)(2) - Metal Area Requirements for Cable Trays Used as Equipment Grounding Conductors,** is reproduced as **Fig. 11.3** in this book.

Table 8 in Chapter 9 of the NEC contains information that is needed to convert AWG sizes of conductors to their equivalent areas in circular mils. This is needed to determine the equivalent sizes of paralleled conductors. This table is partially reproduced in **Fig. 11.4** in this book.

SEC. 250-66

The basic rule in sizing grounding electrode conductors is to simply look them up in Table 250-66. If you're connecting to a water pipe electrode, or to effectively grounded building steel, or to an other metal un-

Maximum Fuse Ampere Rating, Circuit Breaker Ampere Trip Setting, or Circuit Breaker Protective Relay Ampere Trip Setting for Ground-Fault Protection of Any Cable Circuit in the Cable Tray System	Minimum Cross-Sectional Area of Metal (in Square Inches)	
	Steel Cable Trays	Aluminum Cable Trays
60	0.20	0.20
100	0.40	0.20
200	0.70	0.20
400	1.00	0.40
600	1.50	0.40
1000	—	0.60
1200	—	1.00
1600	—	1.50
2000	—	2.00

Fig. 11.3 Metal area requirements for cable trays used as equipment grounding conductors. Derived from NEC Table 318-7 (b)(2).

derground structure, such as some well casings, you'll find the size directly from the table. However, there are some important qualifications to this procedure.

(a) If you are connecting to a rod or a pipe electrode, or to a plate electrode, the portion of the grounding electrode conductor that runs solely to one of these made electrodes need not be larger than No. 6 (or No. 4 Al). This is because the resistance of these electrodes is so inherently high that a sustained current above the short-time carrying capacity of a No. 6 is almost impossible.

(b) If you are connecting to a concrete-encased electrode, the portion of the grounding electrode conductor that runs solely to one of these electrodes need not be larger than No. 4.

(c) If you are connecting to a ground ring electrode, the portion of the grounding electrode conductor that runs solely to one of these electrodes need not be larger than the size of the conductor used in the ground ring. Sec. 250-50(d) sets a minimum size on such conductors as no smaller than No. 2.

In some cases you may need to go in one direction for half the elec-

Size (Avg.)	Area (Cir. Mills)
18	1620
16	2580
14	4110
12	6530
10	10380
8	16510
6	26240
4	41740
3	52620
2	66360
1	83690
1/0	105600
2/0	133100
3/0	167800
4/0	211600
250	—

Fig. 11.4 Circular mil areas of AWG-sized conductors. Derived from Table 8 of Chapter 9.

trodes, and then back the other way for the rest. The Code allows you (in Sec. 250-50) to set out multiple grounding electrode conductor runs from the service. In this case, size each conductor based on the largest requirement applicable. For example, if the system requires 1/0, and one run goes to a water pipe, and the other run goes to a ground ring and a ground rod, make the first conductor a No. 1/0, and the second a No. 2. Then, go from there (the ground ring connection) with a No. 6 to the rod electrode. The water pipe requires a full-sized electrode.

Following are the various sections of Article 250 that require sizing of the grounded and grounding conductors. Because some calculations closely relate to different sections in Art. 250, we are grouping these examples based on related calculation procedures.

Sec. 250-20(d) Separately derived systems. FPN No. 2 deals with systems that are *not* separately derived and are not required to be grounded as specified in Sec. 250-30. FPN No. 1 gives as an example ac on-site generators with their neutral solidly interconnected to a service-supplied system neutral. In these cases, the minimum size of the conductors that must carry fault current are to be determined per Sec. 445-5.

Article 445 gives the code rules for generators. In Sec 445-5, it says that conductors that must carry ground-fault currents must not be smaller than required by Sec. 250-24(b) which, in turn, deals principally with the rules for sizing the grounded (neutral) conductor.

• It must not be smaller than the required grounding electrode conductor specified in Table 250-66, but is not required to be larger than the largest ungrounded service entrance phase conductor.

• For service-entrance conductors larger than 1100 kcmil copper (or 1750 kcmil aluminum) the grounded conductor must not be smaller than 12.5% of the area of the largest service-entrance phase conductor.

• Where the service-entrance phase conductors are paralleled, the size of the grounded conductor must be based on the equivalent area for parallel conductors.

EXAMPLE: As shown in **Fig. 11.5**, a 500 kW engine generator has its neutral solidly tied to the service neutral and is, thus, not a separately derived system. If the ungrounded phase conductors to the load are made up of two 500 kcmil copper conductors per phase, what is the minimum size of neutral conductor?

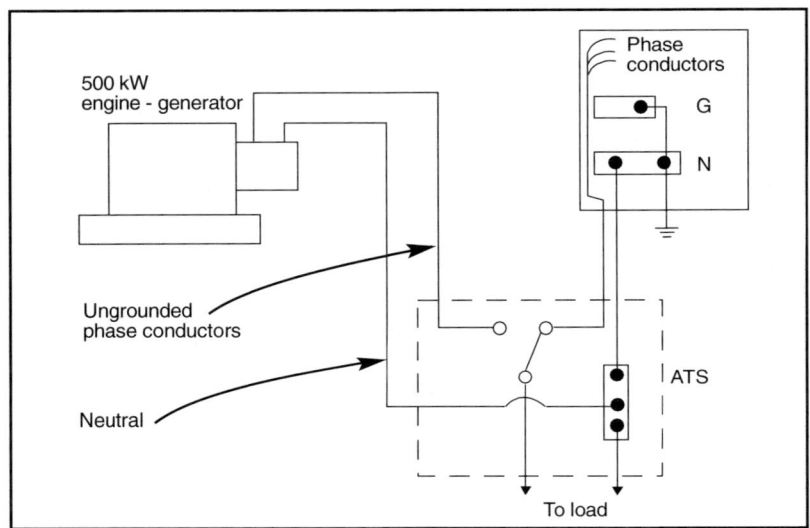

Fig. 11.5 Sizing neutral of a generator not considered a separately derived system.

ANSWER: According to the note following Table 250-66 (**Fig. 11.1**), where multiple sets of cables are used, the size of the phase conductors is determined by the sum of the areas of the conductors.

Table 8 in Chapter [**Fig. 11.4**] gives the areas of conductors in circular mils. In this problem, the size of the conductors is already in kcmils. The equivalent conductor is:

2 x 500 = 1000 kcmil

Per Table 250-66, required minimum size of the neutral is No. 2/0 copper.

Sec. 250-28(d). Sizing a main bonding jumper. These jumpers start out being based on Table 250-66, but the conductor sizes continue to increase because although even the best grounding electrode has appreciable resistance, a main bonding jumper has essentially no resistance; it carries the full value of any fault current back to the grounded circuit conductor. Therefore, the main bonding jumper, for applications not covered in Table 250-66 (i.e., above 1100 kcmil copper or 1750 kcmil Al), must not be smaller than 12.5% of the cross-sectional area of the largest ungrounded phase conductor. Paralleled phase conductors are con-

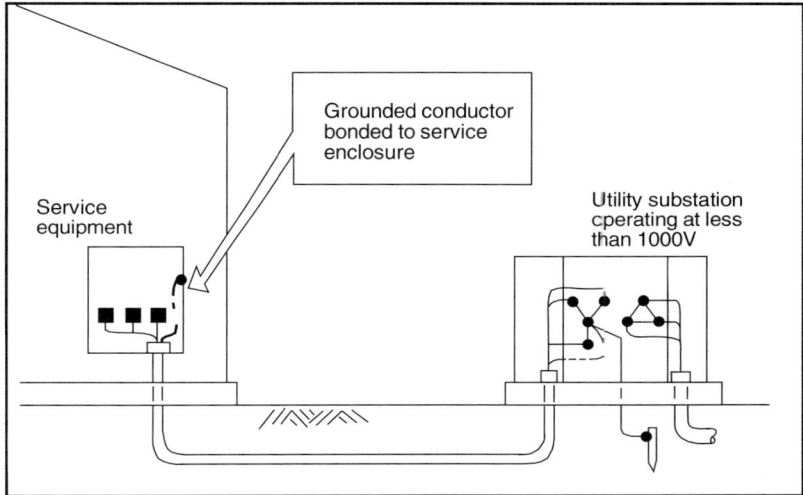

Fig 11.6 Sizing grounded conductor from system operating at less than 1000V to service equipment enclosure.

sidered on the basis of the total area of the components of the largest phase leg.

EXAMPLE: As shown in **Fig. 11.6,** the secondary of a 1500 kVA utility transformer provides 480V, 3-phase, 4-wire service to a building. The transformer secondary current is 1800A. Assuming that the ungrounded conductors consist of four sets of 750 kcmil copper conductors per phase (1900A), what size main bonding jumper is required to be run from the neutral to the grounding bus in the service equipment?

ANSWER: According to the notes to Table 250-66, (**Fig. 11.1**), the size of the paralleled service-entrance cables is determined by adding the areas of the individual cables. The calculation is:

4 x 750 kcmil = 3000 kcmil

According to the table, a No. 3/0 cable is required for a grounding electrode conductor to an electrode that takes a full-sized grounding electrode conductor. For conductors over 1100 kcmil, however, the rule requires that it be larger. sized at not less than 12.5% of the area of the service-entrance conductor. The calculation is:

3000 kcmil x 0.125 = 375 kcmil.

The next larger standard size is a 400 kcmil conductor.

Sec. 250-102(c). Sizing equipment bonding jumpers on the supply side of the service. This procedure is the same as for the main bonding jumper, but with two exceptions. First, when you bond the grounding electrode conductor raceway (or cable armor), the jumper need not be larger than the enclosed grounding conductor. Second, when you bond across parallel raceways, you run the bonding conductors in parallel, based on the size that would be required by Table 250-66, etc., considering the largest phase conductor in each raceway by itself.

EXAMPLE: In the above example, how large are the bonding jumpers running from the service equipment through rigid nonmetallic conduit (as shown in **Fig. 11.7**) to an upstream CT cabinet? Assume the service is ungrounded delta.

ANSWER: Now there isn't a neutral or other grounded circuit con-

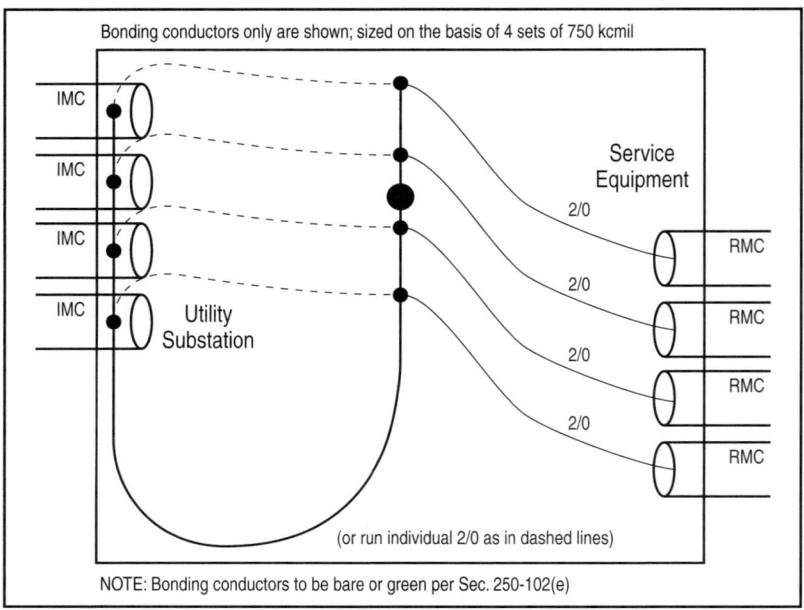

Fig. 11.7 Bonding conductors on the supply side of the service disconnecting means. Where used for bonding purposes only, and not as grounded circuit conductors, the 1/0 minimum in Sec. 310-4 does not apply.

ductor available for bonding, so we have to use additional bonding conductors. Each conduit has a set of 750 kcmil conductors, and that size, per Table 250-66, calls for a No. 2/0 copper bonding conductor in each conduit.

EXAMPLE: Suppose the other side of the CT cabinet has a set of steel conduits going to the utility source. The neat way to bond those four raceways would be to us a single bonding conductor running from bond bushing to bond bushing at the steel conduit side. What size conductor would be required?

ANSWER: Now the calculation is the same as for a main bonding jumper, or in this case, 400 kcmil. Thus, if PVC runs between the service equipment and the cabinet, and IMC runs from there to the utility supply, you could run 2/0 in each PVC conduit, and 400 kcmil from within the CT cabinet to bond from bushing to bushing.

SEC. 250-24–GROUNDING SERVICE-SUPPLIED AC SYSTEMS

(b)(2) Sizing grounded parallel conductors brought to service equipment. Using the same ac system, where an ac system operating at less than 1000V is grounded at any point, the grounded conductor must be run to each service disconnecting means and bonded to each enclosure. This conductor must be routed with the phase conductors.

The following are the rules that apply to sizing this grounded conductor:

• First, as a current-carrying conductor, it must be able to carry the load. In cases where the line-to-neutral load is negligible, however, it still has to meet certain minimum size provisions in order to assure that it will properly perform its other function, of returning fault current.

• It must not be smaller than the required grounding electrode conductor specified in Table 250-66, except that it is not required to be larger than the largest service phase conductor.

• For service-entrance conductors larger than 1100 kcmil copper (or 1750 kcmil aluminum), the grounded conductor must not be smaller than 12.5% of the area of the largest service-entrance phase conductor.

• Where the service-entrance phase conductors are paralleled, all in a single raceway, the size of the grounded conductor must be based on the equivalent area for parallel conductors…BUT

• Where the service-entrance paralleled phase conductors run in separate, paralleled raceways (or cable assemblies), the size of the grounded conductor run in each is based on the largest ungrounded phase conductor run in each, but never smaller than No. 1/0.

EXAMPLE: Going back once again to **Fig. 11.6**, the secondary of a 1500 kVA utility transformer provides 480V, 3-phase, 4-wire service to a building. The transformer secondary current is 1800A. Assuming that the ungrounded conductors consist of four sets of 750 kcmil copper conductors per phase (1900A), what size neutral is required to be run from the transformer neutral point and bonded to the service enclosure?

ANSWER: This is done the same as for the bonding conductors calculated under Sec. 250-102(c) previously. Looking at the 750 kcmil phase conductor in each raceway and applying Table 250-66, you need a No. 2/0 minimum neutral conductor in each conduit. Had the conductors all run in a common raceway, then a single 400 kcmil conductor would have been correct.

EXAMPLE: As shown in **Fig. 11.8,** a building having its disconnect means in another building has an equipment grounding conductor run with the supply conductors. Assuming that a 60A single-phase feeder to the sub-fed building is involved, what size equipment grounding conductor must be run to the sub-fed building? What happens if the grounded circuit conductor is regrounded at the second buildings, per Sec. 250-32(b)(2)?

ANSWER. The answer is found in **Sec. 250-32(f) - Two or more buildings or structures supplied from a common service, as follows: (f) Grounding Conductor.** The set of rules covered by Sec. 250-32(f) gives the requirements for equipment grounding conductors of multi-building setups. The sizing of these conductors is covered by the following rule:

The size of the grounding conductor to the electrodes must, as in the prior example, be able to carry the load. In addition, not be less than the size given in Table 250-122, but is not required to be larger than the largest phase conductor supplying power to the second building.

Note that the size is to be taken from Table 250-122 (equipment grounding conductor table) rather than from Table 250-66 (Grounding Electrode Table). Therefore, a 60A overcurrent device would protect the wires, and, according to Table 250-122 (**Fig. 11.2**) No. 10 copper or No. 8 aluminum wire would be required. This wire size is not larger than the supply conductors, and thus, must be used.

If the neutral is regrounded, and if there is negligible load, it still has to meet minimum size constraints as a fault-return conductor. The minimum size is in Table 250-122, and thus, again, a No. 10 would be the smallest size allowed.

EXAMPLE: As shown in **Fig. 11.9,** a 25 kVA single-phase transformer with a 480V primary and 120/240V secondary feeds a panelboard. What size bonding jumper and grounding electrode conductor are required?

ANSWER: The secondary of the transformer is rated at 104A. In accordance with Sec. 450-3(b), the overcurrent device in the transformer secondary is a 225A frame, 125A trip circuit breaker. No. 1 conductors are selected for the secondary. This will be a lighting and appliance

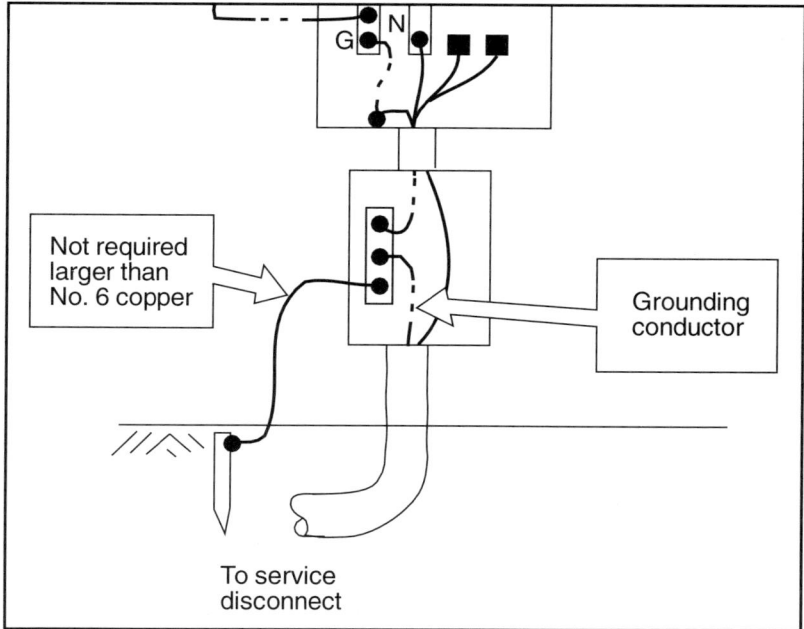

Fig. 11.8 Sizing the grounding conductor in a multi-building complex.

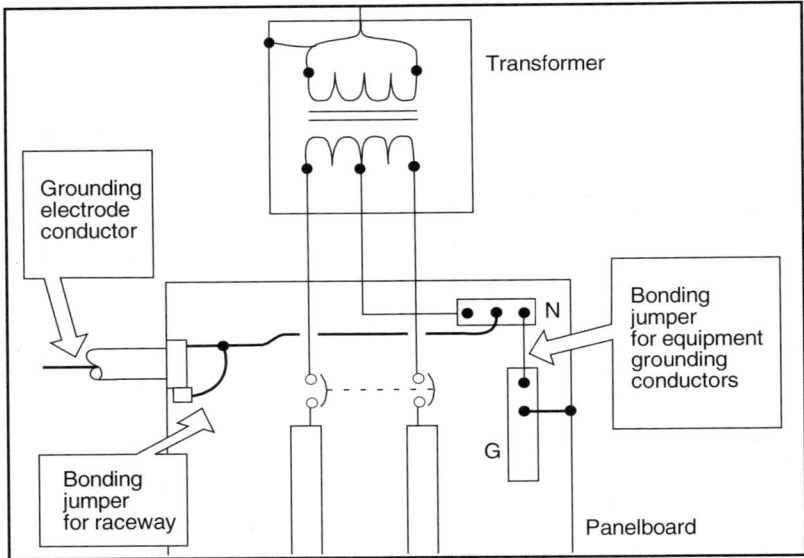

Fig. 11.9 Sizing the bonding jumpers and grounding electrode conductor for a separately derived system.

branch-circuit panelboard with 125A rated bus, so this protection also complies with Sec. 384-16(a).

From Table 250-66 (**Fig. 11.1**), the bonding jumper (based on the No. 1 conductors used) is given as No. 6 copper (or No. 4 aluminum).

Also from Table 250-66, the grounding electrode conductor must be sized as No. 6 copper (or No. 4 aluminum). None of the provisions for certain classes of electrodes would reduce the size required.

EXAMPLE: As shown in **Fig. 11.10,** a 3-phase, delta-connected transformer provides 480V to a distribution system within a facility. The service capacity is 6000A.

Assuming 1A capacitive charging current per 2 MVA of capacity (rule of thumb), after looking at about 6000A x 480V x 1.73≈5 MVA capacity, the charging current would be ≈2½A. The resistor would be set to allow just over this, on the order of 3A, By Ohm's Law, R = E/I, so R = 277V ÷ 3A≈90 ohms. The resistor duty would be based on W = I²R, so W = 3²A² x 90Ω = 810 W, continuous. The continuous rating follows

from the fact that these systems are designed to allow for orderly shut-
downs, so the first fault, should it occur, could send low-level current
into the resistor indefinitely. A No. 14 conductor would have adequate
ampacity, but Sec. 250-36(b), for mechanical reasons, won't allow a neutral conductor for these systems smaller than No. 8 (or No. 6 Al), and
that's the smallest permitted conductor.

EXAMPLE: As shown in **Fig. 11.11,** a 480V, 3-phase, 4-wire service
is brought into the service equipment by three 500 kcmil phase conductors and one No. 1/0 grounded conductor. All the conductors are alumi-

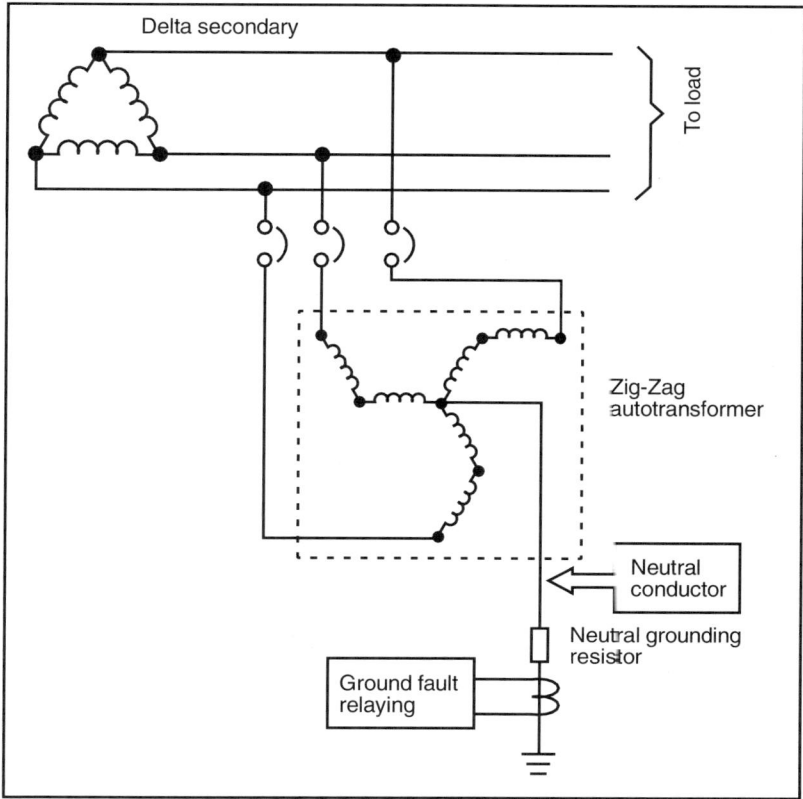

Fig 11.10 Sizing the neutral conductor to a resistor of a high impedance grounded
neutral system.

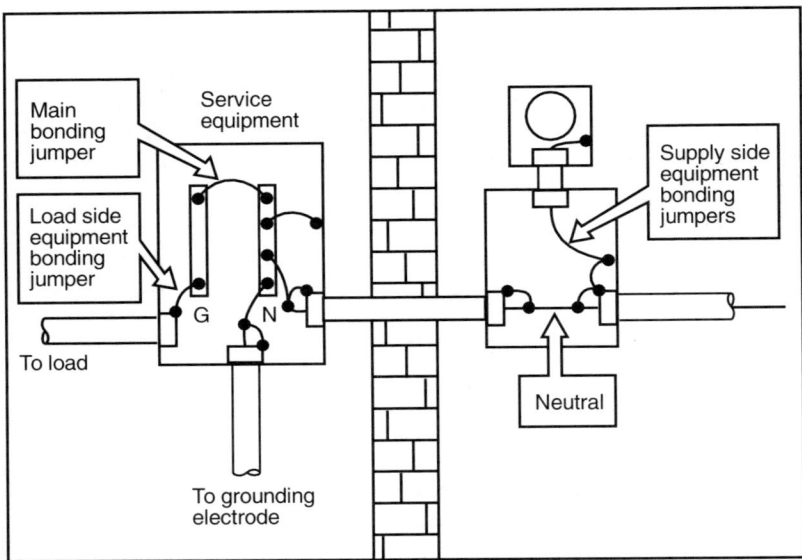

Fig. 11.11 Sizing the bonding jumpers on the supply end and load-side end service equipment.

num, but the bonding jumpers are to be of copper. What size supply-side bonding jumpers are required?

ANSWER: Because of the difference in material, it must be assumed that the service conductors consist of copper conductors with an ampacity of 310A. This is the equivalent to 350 kcmil copper conductors.

The minimum size of the required supply-side copper bonding jumpers shown in Table 250-66 (**Fig. 11.1**) is No. 2. The same result would be obtained if the aluminum conductor in the table was used directly.

Since the table is also to be used to determine the size of the grounding electrode conductor, a No. 2 copper conductor would also be used for that purpose. This assumes that the grounding electrode is one of the ones listed in Sec. 250-50(a) or (b), or Sec. 250-52(b). If, however, the grounding electrode is one of the "made" types listed in Sec. 250-52(c) or (d), then the size of the grounding electrode conductor can be reduced per Sec. 250-66(a) to No. 6 copper.

EXAMPLE: As shown in **Fig. 11.12**, a metallic water pipe enters a multi-occupancy building. At the basement level, the pipe is electrically continuous, but at each other level, the pipe riser isolates the piping within each occupancy. If the service conductors consists of two No. 2 and a No. 8 neutral, what size binding jumpers are required?

If the occupancy subfeed is 60A, single-phase, 3-wire, what size bonding jumper is required to connect the water piping to the occupancy panelboard?

Some miscellaneous metallic piping on the basement is close to a 20A single-phase, 2-wire circuit feeding a water heater. What is the bonding jumper size required to connect the piping to the service panelboard?

If the water heater is changed to gas, what size bonding conductor must be used to pick up the main interior pipe run?

ANSWER: The basement bonding jumper to connect the metal water piping to the panelboard is sized by Sec. 250-104(a) and Table 250-66 (**Fig. 11.1**) at No. 8 copper.

The bonding jumper required in the occupancy, based on wires protected at 60A, is given by Sec. 250-104(a)(2) and shown in Table 250-122 (**Fig. 11.2**) as No. 10 copper. Remember that the same rules apply here as for grounding electrode conductors in terms of installation re-

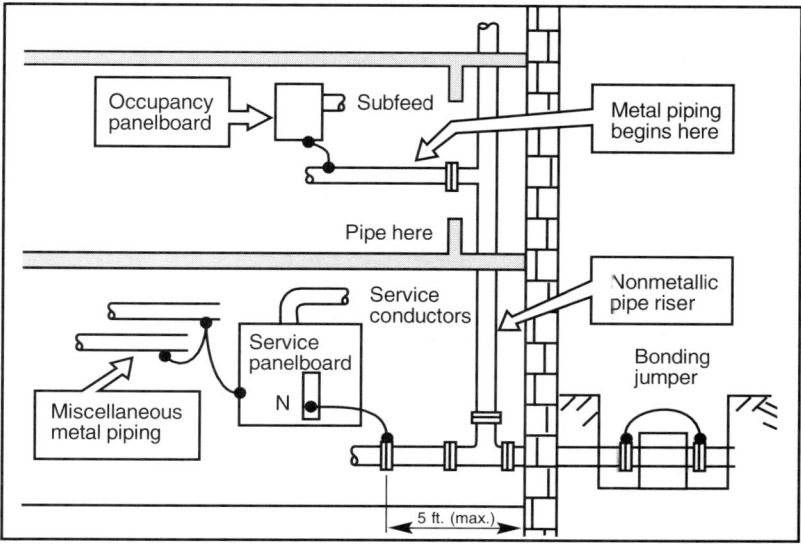

Fig. 11.12 Sizing the bonding jumpers for piping systems.

quirements. The No. 10 bonding conductor must be run in a raceway or cable armor.

The other piping in the basement must be connected to the service panelboard by a bonding jumper sized per Table 250-122 at No. 12. This bonding conductor can be run with the circuit and in fact, it can be the equipment grounding conductor that would run with the circuit anyway.

The gas pipe bond would be sized, conservatively, based on the service size taken through Table 250-66, or No. 8. That size bonding conductor would have to run in raceway or cable armor, so a No. 6 might be used instead.

EXAMPLE: As shown in **Fig. 11.13,** three copper 350 kcmil phase conductors plus one No. 1/0 neutral are brought into a wireway where they are tapped and feed individual 2-pole, 50A service-entrance circuit breakers. What size grounding electrode conductor and taps are required?

ANSWER: The grounding electrode conductor must be sized at No. 2 copper, per Table 250-66. The taps, however, need only be sized per the largest conductors entering the circuit breaker enclosures. Assuming

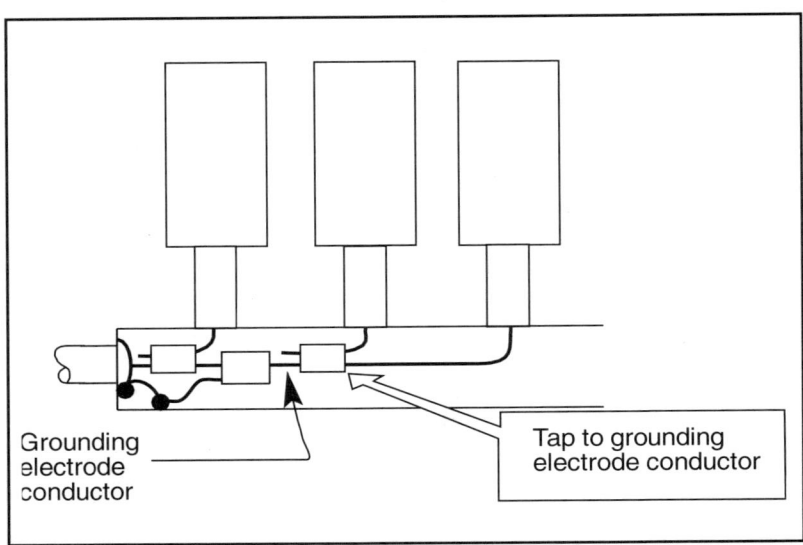

Fig. 11.13 Sizing the grounding electrode conductor and taps where the service-entrance consist of more than one unit.

No. 8 conductors are used for this purpose, Table 250-66 (Fig. 12.1) requires that the grounding electrode conductor taps be No. 8. The No. 2 must be continuous, however, to at least one of the enclosures.

DC systems, Sec. 250-166. The size of dc grounding conductors is covered as follows:

(a) Not smaller than the neutral conductor. Where the dc system consists of a 3-wire balancer set (**Fig. 11.14**) or a balancer winding with overcurrent protection, the Code refers to Sec. 445-4(d), which describes the overcurrent protection needed for a 2-wire dc generator used in conjunction with a balancer set to obtain a neutral for 3-wire systems. In these cases, the grounding conductor must not be smaller than the neutral conductor. In addition, this conductor must never be smaller than No. 8 copper or No. 6 aluminum.

(b) Not smaller than the largest conductor. Where the dc system is other than derived from a 3-wire balancer set, the grounding conductor must not be smaller than the largest conductor supplied by the system. In addition, this conductor must never be smaller than No. 8 copper or No. 6 aluminum.

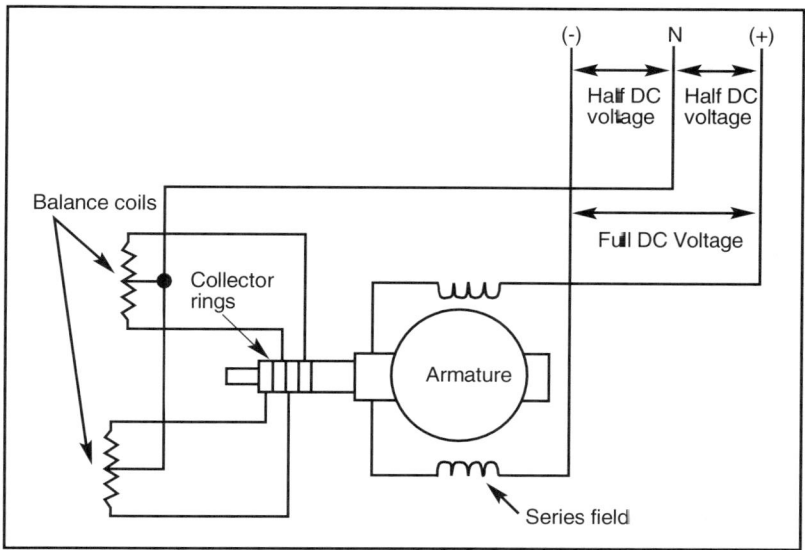

Fig. 11.14 A DC generator with a 3-wire balancer set to derive a neutral.

(c) To specialized electrodes. If the grounding conductor is the sole connection to a rod, pipe or plate electrode, it doesn't have to be larger than No. 6 (or No. 4 Al.) If it is the sole connection to a concrete encased electrode, it doesn't have to be larger than No. 4, and if the same to a ground ring, it doesn't have to be larger than the conductor in the ring. These rules are identical to equivalent provisions for ac conductors.

(d) Bonding jumper (Sec. 250-168). The dc bonding jumper follows the minimum size constraints of Sec. 250-166, just discussed.

(e) Ungrounded dc separately derived systems (Sec. 250-169). The size grounding conductor to the grounding electrode on these systems follows Sec. 250-166, also just discussed.

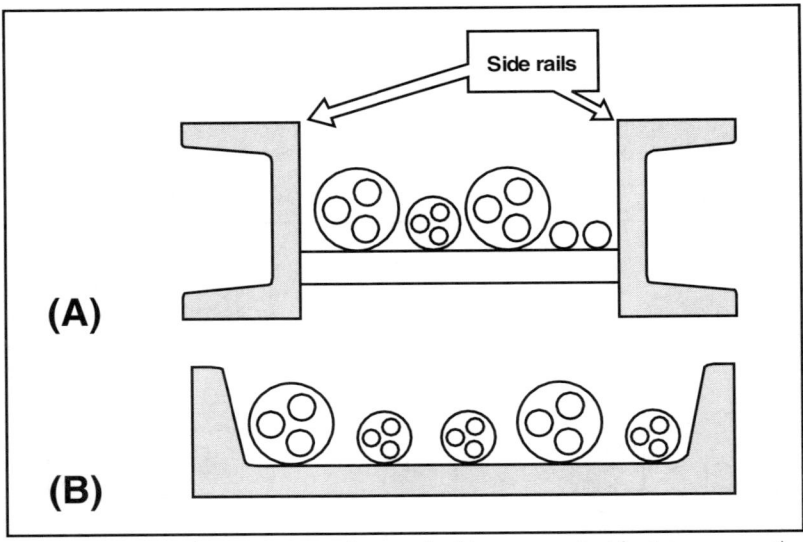

Fig. 11.15 Sizing the cross-sectional area of cable tray required for it to serve as the equipment grounding conductor.

CABLE TRAY EXAMPLE: As shown in **Fig 11.15,** a ladder-type cable tray supports power cables, the largest one of which is protected by fuses rated at 400A. What is the minimum cross-sectional area of metal for: (A) an aluminum ladder type cable tray, and for (B) a steel channel-type cable tray that would allow the trays to act as an equipment grounding conductor, assuming for both that the largest conductors are protected by 400A fuses?

ANSWER: Sec. 250-118(12) lists cable tray among the acceptable equipment grounding conductors, as permitted in Sections 318-3(c) and 318-7. Both steel and aluminum cable trays are permitted by these sections to serve as equipment grounding conductors provided the cable tray sections and fittings are identified for the purpose. Also, all cable tray sections and fittings must be legibly and durably marked to show the cross-sectional area of metal in channel cable trays or other trays of one-piece construction, and the total cross-sectional area of both side rails for ladder or trough cable trays. The following rules apply for sizing the system for equipment grounding purposes. The minimum cross-sectional area of cable trays must conform to the requirements shown in Table 318-7(b)(2). This table is reproduced in **Fig. 11.3.**

From Table 318-7(b)(2), the aluminum cable tray side rails would be required to have a minimum cross-sectional area of 0.40 square inches (258 sq. mm.)

From the same table, the steel tray would be required to have a minimum cross-sectional area of 1.00 square inch (645 sq. mm.).

SEC. 250-122–SIZING EQUIPMENT GROUNDING CONDUCTORS

Refer to Table 250-122, **Fig. 11.2.**

(a) Copper, aluminum, or copper-clad aluminum equipment grounding conductors of the wire type must not be smaller than shown in Table 250-122, but shall not be required to be larger than the circuit conductors supplying the equipment. Where raceways and cable armors are used as equipment grounding conductors, the completed installation must reliably meet the general rule in Sec. 250-2(d). Note that the table itself now refers, with mandatory language, to Sec. 250-2(d) as well.

The performance of the equipment grounding return path is paramount, and the wording has been strengthened accordingly. Under some adverse conditions such as very long runs, you might be forced to further increase the size of a grounding conductor, for example.

(b) **Adjusted for voltage drop.** Where phase conductors are increased in size to compensate for voltage drop, equipment grounding conductors must be proportionately increased according to circular mil conductor area.

EXAMPLE: A 100A circuit running 250 ft was increased two wire sizes, to No. 1 from No. 3. What is the minimum size equipment grounding conductor to be installed?

ANSWER: The area of a No. 3 is 52,620 cm, and No. 1 is 83,690 cm. This is 1.59 times the size of the smaller conductor. Table 250-122 requires a No. 8 to run with 100A circuits, with an area of 16510 cm. Multiplying that by 1.59 gives a conductor just a hair larger than a No. 6, so by the literal text, the equipment grounding conductor would need to be a No. 4.

(c) Multiple circuits. One equipment grounding conductor can be used for multiple circuits in the same raceway or cable if it is sized for the largest overcurrent protective device protecting conductors in the raceway or cable.

(d) Motor circuits. If a motor circuit originates at an instantaneous trip circuit breaker, then the equipment grounding conductor can be based upon the ampere rating of the motor overload device instead of on the instantaneous trip rating of the circuit breaker.

(e) Flexible cords and fixture wire. Instead of complying with item #a above, equipment grounding conductors that are part of flexible cords or used with fixture wires in accordance with Section 240-4 are permitted to be not smaller than No. 18 copper.

(f) Conductors in parallel. The normal rule for parallel circuits is that the full size equipment grounding conductor runs in every conduit. This is because there are conditions under which one equipment grounding conductor will return almost all of the fault current, current that will back-feed into the fault on the other paralleled conductors through the common terminals. In the case of raceway installations this isn't a problem, but cable assemblies usually have an equipment grounding conductor appropriate to the likely overcurrent protection ahead of the ungrounded conductors.

This has meant that cable assemblies for parallel circuits usually needed to be special ordered with oversized grounding conductors, which is a burden. Now, however, you can use Table 250-122 to size equipment grounding conductors in parallel cable assemblies based on the trip set-

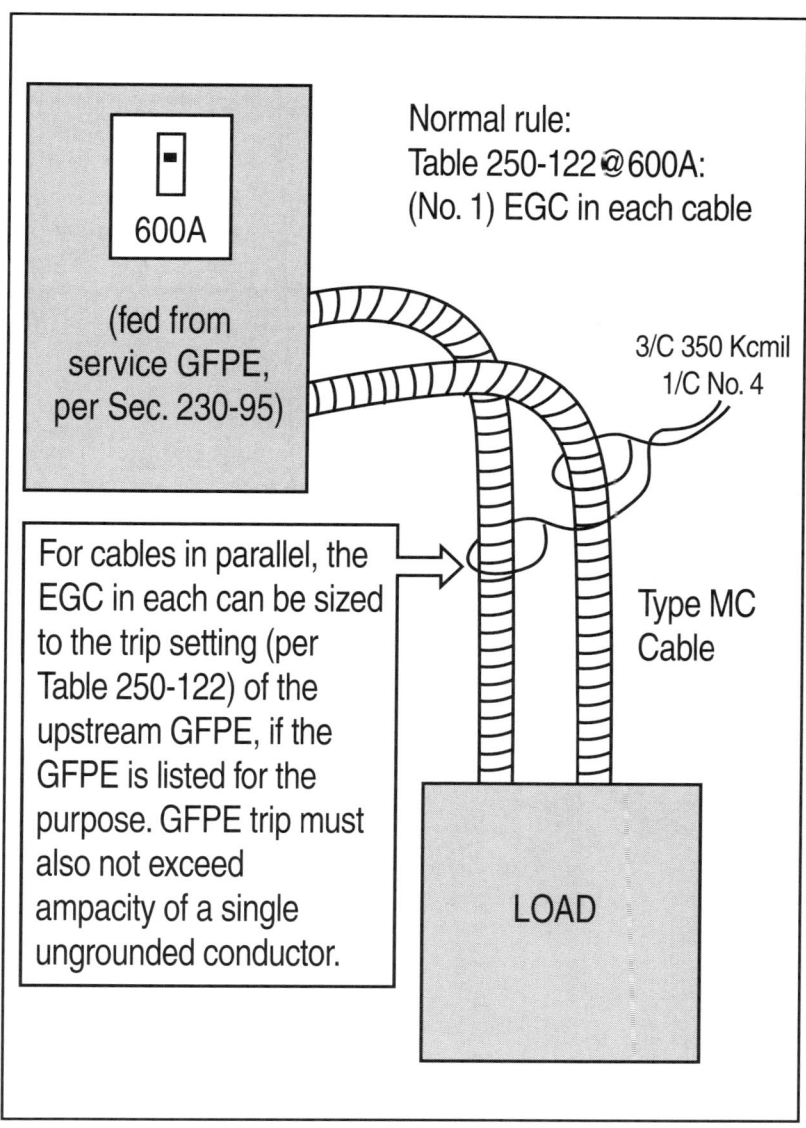

600A

(fed from service GFPE, per Sec. 230-95)

Normal rule:
Table 250-122 @ 600A:
(No. 1) EGC in each cable

3/C 350 Kcmil
1/C No. 4

For cables in parallel, the EGC in each can be sized to the trip setting (per Table 250-122) of the upstream GFPE, if the GFPE is listed for the purpose. GFPE trip must also not exceed ampacity of a single ungrounded conductor.

Type MC Cable

LOAD

Fig. 11.16 A new procedure is available for sizing equipment grounding conductors in cable assemblies.

ting of the GFPE, provided three things are true.

- First, there must be qualified maintenance and supervision.
- Second, the trip setting of the GFPE must not exceed the ampacity of a single ungrounded conductor in any one of the cables run in parallel.
- Third, the GFPE must be listed for this purpose. The panel was told that present GFPE equipment now available would probably prove suitable for this purpose and that the appropriate listing accommodations could be made fairly quickly.

The new change allows for conventional cable assemblies under conditions that assure that the circuit will open before any single grounding conductor would have to carry the monster fault assumed by the main rule. Looking at **Fig. 11.16,** if the service GFPE is set to open at 300A, the cables as shown could be used. The cable ampacity is 310A, above the 300A trip setting, and a No. 4 equipment grounding conductor is appropriate for the same circuit size. If the new allowance were not used, then the cables would need to be made up with No. 1 equipment grounding conductors in each, corresponding to the 600A circuit protection.